Achieving Successful Construction Projects

Whether a construction project turns out to be successful or not has a lot to do with the clarity of the client's objectives and how the client establishes and instils a culture throughout the project team. This book's focus is on defining and exploring those attributes of clients or organisations which enable clear communication, and as a result help ensure the project's success.

For senior construction professionals, this book explains how to approach key aspects of projects so that the client's expectations can be anticipated and understood. It also provides information on how other parties can positively influence the outcome of the project and interact with their fellow stakeholders.

Commentaries on real-life projects illustrate how this is achieved in practice, and common pitfalls are pointed out to help you avoid them. Drawing on almost 40 years' UK and international experience of working on major construction projects in a variety of roles, the author provides clear insight into how to efficiently progress a project from inception through to completion. This is hugely valuable reading for client senior decision-makers, project managers, programme managers, design and construction leaders, and those studying all of these subjects.

Ian Gardner is a Director of Ove Arup & Partners Ltd, a member of the firm's UKMEA Region Board, and has chaired their Future Business Executive. Over his 38 years with the company, he has worked in a variety of leading roles on projects such as High Speed 1, the London 2012 Olympic Park, and the regeneration of Potsdamer Platz, Berlin and central Doha, Qatar.

Ian Gardner and I worked together on the influential Glaxo Group Research project at Stevenage, which in its time helped to establish a new benchmark for major construction projects in the UK. The author is a highly regarded senior professional engineer and leader for one of the world's top firms. His book is insightful in helping to articulate best practice to achieve successful projects. It picks up on the themes of people and how to get them working together, the effective use of processes and management of risk.

Andrew Wolstenholme, CEO, Crossrail

An authoritative perspective from an author with extensive experience and professional expertise in project administration and construction supervision. From project concept, through inception, procurement to completion and hand over, this book packs a wealth of information into its pages. A must read for all engaged in project management, not least the client, architect, site manager and specialist consultants.

Roger Greeno, author of Building Construction Handbook

When many of us are asked what makes a construction project successful, we often say it must be interesting and challenging, the parties must be enthusiastic, competent and realistic, but in many cases these thoughts are subjective, often laced with an element of nostalgia for a favourite past project. The author, who is an experienced practitioner, has objectively encapsulated the ingredients which make a project successful, including project definition, the role of the client and other parties, the formation, and interaction of teams, all referenced to real projects, and all written in a very readable style. A thoroughly enjoyable and insightful read!

Kelvin Hughes, author of Understanding NEC3: Engineering and Construction Short Contract

Ian Gardner speaks in this important book from a position of consummate experience, knowledge and wisdom. The construction industry needs such an impressive ambassador and he makes the compelling case for a better understanding of its challenges; increased efficiency, predictable outcomes and outstanding safety – all possible following Ian Gardner's impressive work. Society needs better infrastructure, of which construction is a major part, and everywhere there are calls for increased investment. Clients, funders and deliverers of construction should all read it.

Terry Hill, Trustee, Arup Group Trusts

A refreshingly easy to read and insightful treatment of major project delivery, with a focus on influence the organisational and people aspects of the equation, especially on the client side, have on successful delivery.

The book will be useful for current and aspiring 'competent clients', and the case studies in particular show clear lessons and differentiate between the good, and the others.

Ian Galloway, Director, Capital Delivery, National Grid

This book examines how a project manager can help a client to define what they actually want to achieve. In addition the author suggests that levels of client involvement, risk and other issues should be agreed at a very early stage.

Clear, well-written and insightful, this book will be warmly welcomed.

Geoff Reiss, Honorary Fellow of the Association for Project Management

Achieving Successful Construction Projects

A guide for industry leaders and programme managers

Ian Gardner

Routledge
Taylor & Francis Group

LONDON AND NEW YORK

First published 2015
by Routledge
2 Park Square, Milton Park, Abingdon, Oxon OX14 4RN

and by Routledge
711 Third Avenue, New York, NY 10017

Routledge is an imprint of the Taylor & Francis Group, an informa business

© 2015 Ian Gardner

British Library Cataloguing-in-Publication Data
A catalogue record for this book is available from the British Library

Library of Congress Cataloging in Publication Data
A catalog record has been applied for

ISBN: 978-1-138-82138-5 (pbk)
ISBN: 978-1-315-74334-9 (ebk)

Typeset in Goudy
by Fakenham Prepress Solutions, Fakenham, Norfolk NR21 8NN

Contents

About the author

Ian Gardner is a Director of Ove Arup & Partners Ltd, a major international consulting group and the creative force at the heart of many of the world's most prominent projects in the built environment and across construction. Ian is a member of the firm's UKMEA Region Board and leads their global activities in Energy. He has chaired the UKMEA Future Business Executive as a member of the Arup Group Business Executive.

He graduated in Civil Engineering from Imperial College London in 1976 and joined one of Arup's Building teams to work with internationally acclaimed architects on award-winning projects.

Over his 38 years with the company, he has worked in a variety of leading roles on projects such as High Speed 1, the London 2012 Olympic Park, Glaxo Stevenage and Qatar Urban Regeneration in Doha. He has been involved in international projects in Australia, USA, Spain, Germany, Belgium and the Middle East.

Between 2008 and 2011 he formed and was leader of Infrastructure London – a group of 400 staff covering transport, energy, heavy civils, site regeneration, marine, geotechnics, tunnels and bridges. He has considerable experience in the leadership and management of large multi-disciplinary design and construction teams, and in all stages of projects from inception, design, consenting, procurement and construction through to completion.

Between 1996 and 2005 he was in lead roles on the CTRL (High Speed 1) rail project, setting up the initial project structure and work plans to establish the 900-strong RLE organisation. He was Project Manager with responsibility for the design, procurement strategy and construction management of the refurbished and expanded St Pancras Station and the major new railway and highway infrastructure in the King's Cross Lands.

He is familiar with developing and delivering major projects on time and budget, whilst also achieving award-winning quality. Key factors in achieving success have been the early definition of the project criteria to meet the client's objectives and clear implementation procedures necessary to control design development and safe construction delivery. He led the setting up of Arup's infrastructure team for the London 2012 Olympic Park.

His international experience includes periods being based in the Sydney office of Ove Arup & Partners Australia and in Germany, where in 1993 he helped to establish Arup GmbH as leader of the newly established Berlin office. He lectured on construction at the Weissensee Architectural College in Berlin in 1995/96. He was a member of an invited senior alumni panel for a centenary review of civil engineering degree courses at Imperial College.

Foreword

The construction sector is of vital importance to the UK's economy. It is a major generator of economic growth and wealth creation, contributing some £90 billion gross value added per year. Major construction projects provide jobs and employment in their delivery – with construction accounting for 10% of the UK's working population. In addition, major projects support and facilitate socio- and economic activity, often in ways that are surprisingly wide ranging and not limited just to the local area. It makes achieving excellence in construction hugely relevant.

Until a couple of decades ago few were familiar with the term 'infrastructure'. The media, politicians and policy makers spoke of roads, railways, power supplies and water networks as separate stand-alone elements. And, most of the population just expected them to be available. Over recent years this has changed and the concept of infrastructure, and even integrated infrastructure, has gained traction and common usage. The adequate performance of the nation's infrastructure is no longer taken for granted. It matters that projects to deliver our infrastructure need to be exceptional and inspirational. Cross-fertilisation of expertise across the whole of our construction sector is increasingly appropriate.

This is recognised by strategies such as the industrial strategy for construction, 'Construction 2025' which is a partnership between industry and the UK Government, and the Mayor of London's 'London 2050 – Bigger and Better'.

In this context this book is enormously relevant and very timely. It is good to see many of the themes and the initiatives of these strategies being described in detail, in a way that illustrates how they can be successfully applied. The book provides an excellent collection of many of the best practices and thinking that help to achieve the desired success. It usefully presents these from the perspective of the practitioner and aims to help those fulfilling the critical client role.

Major construction in UK is already amongst the safest in the world and is fast gaining an international reputation for linking this with reliable delivery. Indeed, for the first time in more than a generation, other countries now look to the UK to learn how to manage their major projects, on time, to budget and with reduced carbon emissions. Spreading this knowledge and expertise is always welcome. It is strategically important for these best skills to be valued and sought after in international construction markets. It is a virtuous circle to benefit from, exchange and grow global influence, while increasing market share, as this also brings international knowledge back to home markets.

Research and innovation is also important to keep the sector competitive and increasingly efficient. Investment in these will not happen without long-term client commitment and confidence that construction is not unduly risky. Clients need to be able to ask for and expect innovation whilst understanding how the correct procurement options help enable this.

It is excellent to see this book contributing to our agenda to maximise the importance of a strong and resilient construction sector.

Peter Hansford
Government Chief Construction Adviser
March 2015

1 Preface
The book's approach

General

This book is aimed at senior decision-makers in organisations undertaking major complex construction projects. It gives them an insight into how to achieve successful out-turn results, by discussing the various aspects of their role as the leader and client. Senior figures running businesses or capital programmes that involve construction projects may or may not associate themselves with the term 'client' – it is used in this book as the designated focal point of responsibility within the commissioning organisation.

This book is also relevant to those responsible for forging strong positive relationships with client organisations, by recognising and responding to the aims of successful clients.

The term 'out-turn' is used throughout the book to mean the end result in terms of performance, achievements and functionality of the completed project.

Large projects or capital programmes are notoriously seen as difficult and complex. They often do not achieve the desired objectives. However, this need not be the case. This book sets out many of the factors common to success. It does so by considering the attributes of leaders and client organisations that have enabled clear thinking. By doing so it provides an understanding of how to achieve the right results with increased confidence.

It recognises that there is no single magic formula or 'one size fits all', and provides explanations of how to approach the key aspects of projects so that the consequences and expectations associated with the main client inputs can be anticipated and understood. In doing so it provides information on how other parties are best able to contribute to a project's success.

The book has wider relevance than just to the leaders of organisations in the client role. It will be of interest to all parties involved in the development and delivery of large projects – designers and management professionals, suppliers and contractors. It is aimed at large projects but much of it can also be applicable to programmes of multiple projects. The book provides a useful understanding of how to take early inception and creative thinking through to a completed project and, as such will be of interest to undergraduates or post-graduates studying Engineering, Architecture, Management or Business courses.

The international nature of the construction industry means that much of the book is directly relevant to and applicable beyond the UK. Other sectors involved in major one-off projects, such as major spending departments in government, the defence, transport, health, international sports, aviation and utilities industries, will also find the book useful reading.

Major spending authorities such as government departments often find it difficult to have confidence in the delivery of large capital programmes and in knowing how to lead them. In the UK, the Treasury has set up Infrastructure UK (IUK), with a focus on cost, improving infrastructure delivery, and the processes associated with project procurement and staged approvals. This book both covers these aspects and provides complementary explanations of other broader ingredients needed for successful outcomes.

Advantages of this book's approach

Much of the thinking and guidance associated with project delivery is linked into the processes and procedures of project management. The implication being that with the right project management, the desired outcomes will be delivered purely through management processes. An extension to this approach is often to have the contractual basis for a project well enough tied up so as to place responsibility not with the client, on a one-sided basis. The premise being that the management processes can simply keep the pressure up on the delivering parties if they are contractually bound to pick up and handle whatever arises.

This book takes a different position. Success cannot be achieved just by passing responsibility contractually to others and by management imposing processes. An enlightened, informed leader in the client organisation is needed as an on-going controlling influence and participant in the project. This book is unusual in discussing and describing the attributes needed on the client side. It does so in a way that means that others can in turn consider how their inputs fit and are likely to contribute. The book proposes that the supply chain and stakeholders' expectations need to be understood, and they need to be motivated to be willing to assist in achieving the objectives.

Many books and published texts focus on project delivery. This takes for granted that it is known what it is that is required to be delivered. This book, by contrast, assists the reader in considering the important development phase of projects and being clear on the interfaces between what is wanted and how to deliver it. It is the combination of these that enables successful projects.

> Note that throughout the book the general use of the term 'he' should be read interchangeably as 'he' or 'she'. The term 'capital programmes' refers to a planned collection of projects, but elsewhere the term 'programme' is used in relation to timeframe, whereas in some countries the term 'schedule' is used in this context. 'Scope' is used as the term for project requirements, whereas the term 'programme' is used for this in countries where 'schedule' is used for timeframe.

Structure of the book

This book is structured as a series of guidelines, each of which is a chapter heading. Using the term 'client' to represent the senior decision-makers in the commissioning organisation, it sets out to complete the statement:

'The most successful clients know .....'

and it does so by discussing and defining this knowledge under the following headings:

The most successful clients know...
  – How to define what is wanted and when
    – How to set up and organise the project team
      – Whether they want to be 'hands-on' or not
        – How to involve the best contractors and SMEs
          – How to best buy the project
            – The relationship they want with the project team
              – How best to align risk, cost and contingency
                – How to measure and control progress
                  – How to approach communication
                    – How to handle the interfaces with stakeholders
                      – The post-completion and legacy objectives

Key to all this is for there to be no ambiguity regarding who is acting in the client role, so this is introduced at the beginning of the book. It is also important for the client to understand which decisions are required when, so the book sets out to link these issues of good project governance to the concept of a project timeline.

Why this book?

This book is largely based on the personal experience of the author. The author has had extensive experience in many aspects of projects, from inception and design through to construction and handover. With the best part of 40 years' experience in the UK and internationally, he is still involved in the leadership of major projects as a senior practicing engineer.

So why write this book?

Over the years the author has given many talks, lectures and presentations on different aspects of projects, and has chaired events relating to the delivery of major projects. He has written articles for publication and was a member of an expert drafting panel for the Institution of Civil Engineers' *Client Best Practice Guide*. In 2007 he was invited by Imperial College to be a member of a small alumni panel undertaking a centenary review of their Civil Engineering degree curriculum and its relevance to the future needs of the profession. And he has contributed to a couple of leading contractors in improving their understanding of and ability to manage design.

This book started as an assembly of these various pieces. Though none were necessarily unique, they were considered worth consolidating. Expression of a broader interest has resulted in these being amplified and transformed into a reasonably consistent format, the result of which is this book. It is not claiming to be an academic reference or a review of related publications. The book is a reasonably authoritative piece of writing based on a desire to describe and help handle the many practical issues associated with significant construction projects.

It is not written in the first person but in many aspects it could have been. It is a reflection of the experience and attitudes that the author sees as most relevant, and describes ways through these that the author believes are most likely to work towards achieving successful projects. The book seeks to provide a bridge from industry guidelines and academic thinking to practical behaviours and awareness that clients and their projects can use.

It does not claim to provide all the answers. Rather, it seeks to be thought-provoking to raise consideration of the key issues that directly influence success. It primes the reader of these issues and the factors associated with each. By doing so, it brings together and embraces many elements of projects which in practice need to be addressed. It is not a long book but the ideas in it cover much of what matters. Taking these on board will increase the likelihood of positive outcomes.

[The diagrams in this book were produced by and remain the copyright of Ove Arup & Partners Ltd. In most cases these diagrams have been created by the author on various projects or for presentations he has given.]

2 The challenge of leadership
Providing a general context

Setting the context

The client is the organisation commissioning the project and paying for it. The role and performance of the senior decision-makers in the client organisation is probably the single most important factor in determining whether or not large construction projects are successful.

This may not generally be recognised and, even when it is, there may be uncertainty over how the client can perform so as to positively influence the out-turn success of their project. Many senior figures in client organisations may feel when a project is less successful that this is the consequence of failings in those they engaged to deliver the project; that they felt powerless to influence things but were left having to accept something they were not satisfied with.

However, on projects where the client has ended up less than satisfied, there is convincing evidence of a high likelihood that many of the other parties involved also found the project unacceptably difficult or unsatisfactory. So there is a tendency for a general 'lose–lose' situation. It is therefore of general interest and benefit to have projects with excellent, high-performing and knowledgeable clients.

This guide for leaders of projects and capital programmes has been assembled both to assist clients directly with the challenges associated with delivering projects successfully and to support construction professionals in providing advice and advocacy to their clients. This guide is therefore written so as to point towards those issues that the client needs to consider and to identify aspects of client performance that help achieve a 'win–win' successful project out-turn. It examines various aspects of the client's input and provides guidance on measures that have been found to be particularly influential.

Clients can be surprised when asked

'*At what point will they know their project will be successful?*'

Either they have not considered this, or do not have a mechanism to plan for success. This guide attempts to illustrate that delivering successful projects need not be difficult.

Leadership or management?

It is interesting to debate whether the issues of *leadership* or *management* are relevant to achieving successful outcomes, and the degree to which they therefore apply to the subject of this book.

Both are clearly needed in order to achieve results.

However, it might be proposed that the greatest success comes from getting the best from as many people as possible.

Leadership at its best is setting a clear vision of the art of the possible that many others can find inspirational and worthy of having the ambition to attain – the *'Why'* and the *'What'*. The leader focuses on a culture of high yet achievable expectations, but not so much on the *'How'*. This then leaves others to feel they are able to truly contribute to the *'How'* through their creativity and engagement. They have the excitement and interest of being able to feel they have *'some skin in the game'*. Provided they are of the right quality, this can lead to high-performing teams that have a common clear focus and can be very committed to achieving results, even on the most challenging of projects. The ambition is clear and, therefore, it is also clear what will represent success.

Management, by comparison, is more about directing others, telling them what to do next and the way to do it. This is more about prescribing the *'How'*. This can be effective when something has been done before and a precedent has been set, so the rules are known and the task is to comply with them in an established way to get the desired result. However, it can be less stimulating and less engaging for those involved and is therefore likely to get less of a contribution from them. If it does not go well, then it does not feel like their fault as they were only doing what they were told, and who were they to argue? If successful then those participating are likely to have a great respect for those making the management decisions but are less likely as participants to feel the buzz of having personally made a difference.

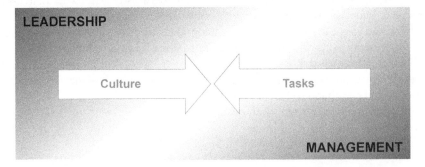

Figure 2.1   Environment for a high performing team – the overlapping zones of Leadership and Management. High performing teams are achieved by leadership focussing on creating the right culture. Management can then focus on the tasks.

A combination of both leadership and management is needed. For the most complex projects to be successful there is a need for a fair measure of inspirational leadership. Really good leadership is clear thinking on the art of the possible, but also generous in giving space and credit to those that make it happen.

The way projects are structured can make a significant difference to whether an atmosphere of leadership or management can prevail.

> *There is an interesting parallel with coaches of top sportsmen. Prior to the London Olympics a gathering of top construction leaders had a meeting with one of the UK's top coaches. Someone asked what coaching had to do with leadership. The response was interesting. The coach explained that top athletes are always afraid of failure. They are happy to agree to challenging targets to stretch their performance ambitions, but then look for excuses for not being able to quite deliver on the day due to factors and influences that were 'outside their control', so not their fault. The best coaches are the ones who can achieve two things. First to get buy-in with their athlete to a belief in the achievability of high targets. Then secondly, and importantly, for the coach to succeed in creating an environment and circumstances such that all the possible excuses for the athlete not performing at their absolute best have been eliminated. The athlete is then mentally in a position where they know nothing is preventing them from achieving their best ever result. They have no excuse. The most effective leaders achieve something similar with their project teams – by making a belief in achieving exceptional outputs the prevailing culture, as there is no excuse for not doing so.*

Having the right support – for development and delivery

Few people in the senior decision-making client role are 'universal man'. Being a leader of successful projects therefore does not mean you have to do it all yourself. As has been explained above, the best leaders do not attempt to do, or control, everything. Understanding what support is needed, and the nature of this support, is consequently important for successful outcomes.

By definition many of the senior figures in client organisations are experts in their particular business or market sector. That is why they are in senior positions leading their business. They may have conceived the capital programme or project as part of a business plan or venture of strategic importance. This will have out-turn objectives and to achieve these they may have identified the associated need for construction works. These construction works will provide new capital assets to take their business or sector forwards. However, the mainstream skills of these senior client decision-makers may not be in construction. Their core role is to drive the business needs and business case.

It does not matter if these senior client figures are not experts in construction if they have the right support in these skills around them; whether within their business or outside it. Given an awareness of the issues, they are able to consider the nature of support required.

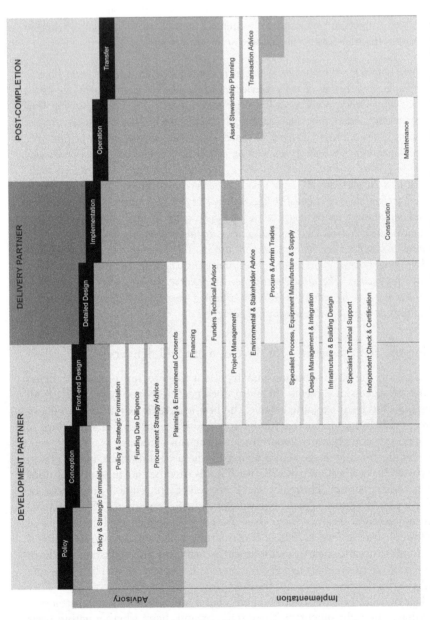

Figure 2.2 Activities associated with the Development and Delivery Phases through to Post-completion, illustrating the split in focus between 'advisory' and 'implementation' inputs.

The 'Why' and the 'What'

- If the support required is to help establish and set up the project and its definition and scoping in order to be clear on what is needed,
- and if there is a requirement for the creative experienced input to establish the means by which it can achieve what is needed in order to deliver out-turn value and meet in-use performance objectives,

then, there is a need for

Project Development Skills (a development partner)

to supplement the client with these skills and expertise.

The 'How'

- If the support required is once the project is clearly defined, in order to help the client with the management of the project execution,
- and if there is a requirement to provide the capacity of experienced resource to achieve this,

then, there is a need for

Project Delivery Skills (a delivery partner)

to supplement the client with these skills and expertise.

It needs to be understood that these roles are different and not interchangeable. Both may be needed, but it is flawed to ask individuals skilled as delivery partners to define the project, and similarly flawed to ask those who specialise as creative development partners to manage delivery.

To help illustrate this it can be useful to consider the activities and roles associated with 'advisory' and 'implementation' inputs.

Historical perspective

It is interesting to note that the significance of the client relationship with other stakeholders in the construction process has long been recognised as a critical factor for success. This has been recognised previously by the Institution of Civil Engineers and in this context the idea of providing guidelines for clients can be traced back at least 250 years, to the beginning of modern civil engineering.

The following example sets out some of the thinking at this time, much of which has relevance today.

John Smeaton, the pre-eminent engineer 1724–1792, and generally regarded as the founder of the civil engineering profession, frequently provided advice to clients on their role. The best known example is contained in Smeaton's correspondence regarding the Grand Canal in Ireland, published as *Letters between Redmond Morres ... and John Smeaton* (1773). Smeaton had been involved with the Forth and Clyde Canal for five years. He had provided their 'Proprietors' with the ideal management structure for that work in 'A plan or model for carrying on the mechanical part of the works of the canal from Forth to Clyde' (March 1768; J Smeaton (1812) *Reports*, 2, 122–124). This set out what the client, or the Canal Directors should expect of the various elements in that structure.

After five years' experience he used a metaphor of watch-making to illustrate the likely technical limitations of a client body: 'were the construction of a watch the subject of direction, it is probable that not one Gentleman on the whole committee can make a watch: the necessity of the thing therefore would oblige them to employ a watch maker, to him they must in a great measure trust. In fact, he must, as to mechanical construction, direct the Directors'. Smeaton recognised that the client might, in other matters 'be competent judges'. For Smeaton the key role of the client was 'to keep a watchful eye over their officers, to see that they are employed in their own departments, and in their own departments only ... the greatest work is to keep right the animal part of the machines; and I will be bound to say; that if the [client's] committees can preserve a good understanding between themselves, and their principal execution offices, and prevent those from falling foul of one another, it will be a saving of a great many thousand pounds to this undertaking, and of much disquiet to those who have the principal management: I have observed that many defects arise from want of will, than from want of skill.' Centuries have passed since Smeaton recognised the need for clear direction and authority, and a cooperative approach, yet these principles underpin good client-stakeholders relations and successful project outcomes to this day.

Another theme touched on by Smeaton, and a continuous theme in client–stakeholder relationships, concerned estimates and costs. The method of procurement was not Smeaton's priority – whether by advertised contract or direct labour / negotiation, his concern was that 'it is not made a condition to prefer the lowest bidder; the merit of the Tradesman, and the goodness of the materials, are in those things of superior consideration to the lowness of the price.' (Smeaton's most notorious failure, Hexham Bridge, was in part caused by his concern to provide an economic solution for his client; a more expensive approach to protect the foundations from scour was not used, and the bridge failed in floods.)

Smeaton's 'Client's guide' represents measured advice based on two decades of experience. Two generations later the engineer John Urpeth Rastrick provided a clarion call to arms for another client, the Directors of the London and Birmingham Railway.

'Let nothing deter you from executing the work in the most substantial manner, and on the most scientific principles so that it may serve as a model for all future railways and become the wonder and admiration of posterity. There is nothing but what a large Splendid Company like yours can accomplish. Remember that faint heart never won fair lady. Therefore let me conclude with the advice of Queen Elizabeth to one of her courtiers: "Climb boldly then."'

Rastrick's advice undoubtedly reflected many engineers' frustration with clients who were undercapitalised, or simply preferred cut-price solutions that brought in their wake shoddy workmanship and incipient failure. Fortunately, the London and Birmingham Directors were conscious of their role in creating a 'prestige project'. Despite escalating costs they supported their engineer and the result was the most financially successful of the main-line railways. The rising costs were those that typically befall clients – the cost of iron rose as there was insufficient capacity in the iron trade of the time. Traffic forecasts were reviewed upwards during construction, necessitating larger facilities. Such experiences may be typical of prototype projects. In this case the immediate success vindicates all involved. Railway construction in the second half of the nineteenth century involved less risk for the client, costs were better known, and projects were generally completed on time and to budget.

3   The client

Responsibility within the commissioning organisation

The client role

Almost all construction projects or major capital programmes have someone or an organisation as the client. This is the organisation that has decided on the need for the project, commissions others to be involved and pays for it. The role of project clients is multifaceted and complex. However, in essence, good project clients focus on the following key tasks:

- Strategy – owning and maintaining the strategy for the project.
- Project culture – creating the right culture and environment for the project and carefully monitoring it, staying vigilant for changes that might impact on the project and its business.
- Business case – establishing a clear business, legal and financial framework at the outset of the project; then constantly revisiting it, verifying its assumptions and its ongoing validity.
- High-level progress – focusing more on prognosis rather than monitoring detailed progress.
- Corrective action – taking clear and timely decisions throughout the design and construction process; being ready to 'press the start button' for corrective action if required; similarly, being prepared to 'press the stop button' if the project becomes unviable or is in need of redirection.
- Communication – communicating widely; to the client organisation, the project team and those affected by the project.
- Stakeholders – seeking to understand and where possible satisfy the requirements of all parties with an interest or concern in the project.
- Lessons gained – learning from other projects and from working closely with other clients, both within the organisation and in the wider industry.

It is in the best interests of a client to have a full understanding at the outset of his requirements, the process to develop a project and what the end goal for the project should be in terms of output delivered (a strategy for the project).

Although it is entirely possible that changes in the client's requirements over the life of the project may lead to amendments to the original brief, the project will benefit from a clear description of what is going to be delivered being made at an early stage, providing confidence to the whole project team.

There are a number of characteristics that define successful project clients. These include having a strong belief in and commitment to the project; setting out the ambition and the art of the possible; being driven by results and committed to success; and holding a genuine concern for the people involved, actively supporting and enthusing them.

> *While it is, of course, the client that will have to take the final decision on any aspect relating to project or programme strategy, he will benefit from taking on board advice from external experienced practitioners in developing this vision. The client may not be able to appreciate issues related to the delivery or end use of a building or structure that will be more obvious to those that have to construct or use them.*

The most effective project clients have a wide breadth of view, and are frequently networking and lobbying. They are thinkers, exploring and challenging issues and angles. They demonstrate courage, taking calculated risks, and show flexibility, adjusting their approach as necessary. They are not unduly concerned with their personal status and, crucially, they are good managers of time, both their own time and that of others. It should be noted that senior managers of a client organisation with operational responsibility may not have the right experience to also make good leaders of construction projects. Generally, these operational managers should stay in the 'client' role of setting requirements and avoid being on the project delivery side of the client team.

The project client

Major construction projects are commissioned by organisations, and these organisations take the client role for the project. At the strategic level the client organisation has to set the agenda and objectives for the project and then usually identifies an individual within its ranks to be vested with the authority and responsibility to be the project sponsor or client. The 'client' for a construction project is thus the individual within the organisation paying for it or procuring it – the 'client organisation' – who is accountable for the benefits of the project to the client organisation's business. Those benefits might be

- Functional – specified purpose and/or improved operation
- Financial – revenue and/or cost efficiencies
- Environmental – carbon reduction, waste or other pollution reduction, climate-change resilience
- Societal
- Reputational
- A combination of the above

Consequently the project client 'owns' the business objectives and required benefits of a project on behalf of the client organisation.

No one is more fundamental to the clarity of a project than an enlightened client.

The organisation commissioning and paying for the project is the client, but senior members of this organisation may have leadership of operational or functional units and thus be the end-user client but not necessarily see themselves as the project client with the hands-on interest or responsibility for its achievement. Multiple representatives in the client organisation might believe they have the right to have their views and requirements taken into account, but they might not see themselves as taking responsibility. In such cases it is important to be clear on who is holding the responsibility for the project in its totality.

Client leadership over the life of the project

The project client plays a vital role at every stage of the project lifecycle: from initial inception through to development, design, construction, commissioning and entry into operation. The intensity of the role changes over the lifecycle, with some stages requiring much greater input than others, but nevertheless the project client has a distinct role to play throughout. Continuity in this role can often be an important factor in successful projects.

The best project clients provide effective leadership of the project at strategic level, with the full backing of the client organisation. This includes creating and communicating a vision for the project, which will enable the business and all other participants to understand its purpose. Effective leadership requires demonstrating personal commitment, typically described by others as – 'This is someone I want to follow.'

Major projects develop their own unique culture – 'The way we will do things around here' – and this stems from the project client. A key element of culture is behaviour, and the behaviour of project clients and their senior team will be closely watched by the rest of the project team. If inspirational, inclusive and collaborative, or if aggressive and adversarial, it will influence the way others respond. Project clients thus have a key influence on behaviour within the whole project team.

Projects typically involve some hard choices so it is necessary to understand and set priorities at the senior client level.

Having a clear client vision – clear about what they want and articulated well.

Everyone involved with projects is familiar with the Time–Cost–Quality triangle. This illustrates the sometimes difficult trade-offs that have to be made between these three primary variables. Good project clients form an understanding at the outset of their priorities for the project in relation to these three variables. This

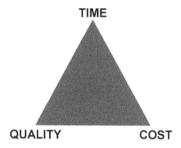

Figure 3.1   Time-Cost-Quality triangle.

does not mean they will be soft on certain objectives but will have a framework established in which to make tough choices if required.

Note that sometimes Safety is considered as a fourth variable. However, this guide does not support the concept of safety being a variable; safety should never be traded or offset in order to achieve other objectives.

Who is in the client role?

There is a tendency on projects to assume that there is only one client, but whilst there may only be one ultimate client many members of the project team may also be fulfilling client roles.

There may be a party (or individual) acting as the agent for the client, with delegated authority to act on behalf of the ultimate client. Other parties involved in the project may see this agent as their effective client.

Parties involved in the project with a direct contract with the ultimate client may themselves appoint sub-consultants or subcontractors, for whom they are in turn fulfilling a client role. They have employed these other parties and so will be seen by them as their client.

So the 'client' is those individuals (or parties) that others see as empowered to fulfil the role of their client.

Therefore these guidelines can apply to all parties fulfilling a client role.

In many projects there are individuals or authorities empowered to give permitting or grant approvals. Often they behave as though they are the client, but they are not, and a better understanding of this distinction is important. In the context of this guide they should be considered as stakeholders with a particular remit to grant acceptance, rather than as the client.

Governance

The term 'governance' is used to describe the way in which the project is authorised, conducted and overseen by the client organisation and significant interested parties, such as sponsors, funders and regulators. It is a mechanism for engaging the client organisation in the project, for securing buy-in of key players and for driving executive decision-making. Hence project governance is one of the most important roles that the client fulfils. Good governance helps to eliminate uncertainty; with both risk and out-turn overruns commonly being a consequence of uncertainty. The governance process crucially includes establishing appropriate and effective delegated levels of responsibility.

> *Ideally project governance should be set in place before contract award, as it will affect how a supplier chooses to price the work. If the governance is not clear from the outset, this will affect supplier confidence in the client as an organisation, and their bid will reflect this. Just as clients may ask suppliers to demonstrate that key staff are committed to the project, the client organisation itself should be able to show that it has committed suitable staff to the project in order to build confidence.*

There is not a 'one size fits all' governance structure; project governance needs to be appropriate to a particular project and client organisation. Client organisations typically establish a project board to fulfil the governance function. Where an organisation carries out multiple projects, a programme board or portfolio board might be established, with project boards reporting to it.

It is essential that the project client, project manager and operator all attend the project board, preferably as full members. Other parties might be represented on the project board, particularly if they have a funding or equity stake in the project or if their involvement is vital to the success of the project. It is normal practice for the project client to chair the project board.

Research has shown that a governance process is at its most effective when all interested parties are actively involved and they can communicate constructively in an open environment.

> *There is nothing worse than going to a designated person only to find that they do not have the power to act. There is a need to identify who the key people are in project governance at the outset of a project.*

Relationships with others

Very few client organisations are able to undertake in-house all roles needed for the development and delivery of a project. The project client therefore has an important duty to assign and mobilise staff from within his organisation and/or appoint organisations who are specialists in order to execute the project. These key relationships to achieve a successful project outcome might include the following:

- Client organisation's internal project team
- Funding advisors
- Legal advisors
- Insurers

- Media and public-relations specialists
- Planning and environment consultants
- Health and safety advisors
- Design team
- Project manager
- Contractor(s)
- Specialist trades, suppliers and manufacturers
- Operator and maintenance team

The term 'supply chain' is commonly used to describe a number of these parties that are external to the client's business – it can be used to refer to all or combinations of the interlinked group of consultants, contractors and manufacturers that provide the various elements of the work during the implementation stage of the project, the input of which is typically managed by the project manager

The client, the project manager, consultants, contractors, the supply chain and the operator need to work as an effective 'project team' for successful achievement of the project.

> *Given the responsibility that most clients pass to their supply chain [often via the project manager] to deliver their projects, making sure that the right organisations are attracted in the first place is one of the most important elements of the client role.*

Avoiding client ambiguity

On large, complex projects it is very important to achieve and identify a singular unified client body to have responsibility and be empowered with the necessary authority to make things happen and get to a result.

Responsible: Has the unambiguous ultimate say with the authority to do so, and responsibility for clearly communicating the agreed requirements and decisions taken as appropriate to all those involved.

Accountable: Has key inputs to the project in terms of setting performance standards, client brief and quality requirements for their parts of the project and are held to account for the timeliness, accuracy, completeness and justifiability of these. Examples would include heads of departments, heads of operation and maintenance, or heads of security. If it turns out that these inputs collectively lead to incompatibilities then these need to work with those responsible to reach an agreed solution.

Support: Not always used but on large projects a peer group, a team of external industry experts, or marketing/communications specialists may be established to provide independent support and advice to help inform the responsible client. These are advisors and not decision makers.

Consulted: Those parties or bodies that the project has a requirement or obligation to seek informed views from. Where possible their views will be taken on board but there is no absolute requirement to do so as they are not the client. It is recommended to maintain an audit trail of these consultations and the reasoning adopted on receiving their views.

Informed: This is the group or interested population who will be kept informed so they have an accurate awareness of the project rather than through ill-informed rumour, but they are not asked for a view.

For the good governance of significant projects it is strongly advisable to set out and get high level approval of a RAC(S)I schedule populated with parties and individuals to identify their remits. This should then be maintained and kept up to date. This avoids unnecessary ambiguity over who is 'calling the shots' and gives recognition to and empowers those who have a role to play.

4 Project timeline

The project stages over its lifetime

Introduction

The challenge for many clients when embarking upon large, complex projects is to achieve an appropriate match between the level of investment and the out-turn functionality and performance of the project. Delivery of the project will need major commitment if it is to be a success, but this leads to three questions:

- What will represent success?
- How will the project be defined?
- Through what process will it be delivered?

In response to these, almost all successful and experienced client organisations recognise that their projects go through a number of phases or workstages. These are structured to enable the viability of the project to be tested and then to enable it to be monitored and controlled as the project progresses. This structure provides a *project timeline* and helps to identify which key decisions and activities are required at which stage. The client organisation needs to understand their inputs to the project in the context of this timeline.

Clients for the most successful projects recognise and assess the following five stages when strategically considering their projects:

- Planning
- Development
- Implementation
- Operation
- Decommissioning

This guide provides guidance to clients through these five stages of the project timeline on those activities that need to be considered at what point in the project.

Figure 4.1   Sequence of five project stages.

Commonly used workstage methodologies

The importance of having planned workstages through to project completion is recognised by many leading construction organisations, including

- Royal Institute of British Architects (RIBA) – Plan of Work stages A–L
- Office of Government Commerce (OGC) – Gateway Approvals 1–5
- Network Rail (NWR) – GRIP stages 1–8
- Transport for London (TfL) – Corporate Gates A–E

These are depicted below against a common basis to enable them to be compared.

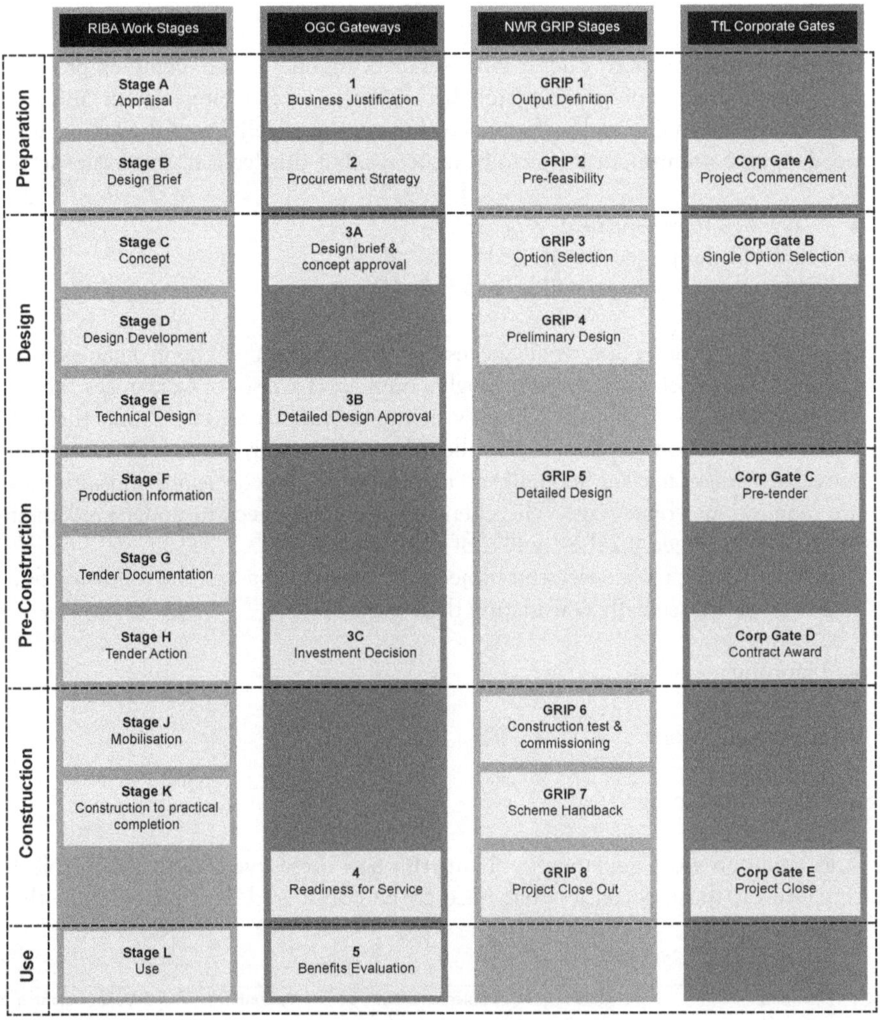

	RIBA Work Stages	OGC Gateways	NWR GRIP Stages	TfL Corporate Gates
Preparation	Stage A Appraisal	1 Business Justification	GRIP 1 Output Definition	
	Stage B Design Brief	2 Procurement Strategy	GRIP 2 Pre-feasibility	Corp Gate A Project Commencement
Design	Stage C Concept	3A Design brief & concept approval	GRIP 3 Option Selection	Corp Gate B Single Option Selection
	Stage D Design Development		GRIP 4 Preliminary Design	
	Stage E Technical Design	3B Detailed Design Approval		
Pre-Construction	Stage F Production Information		GRIP 5 Detailed Design	Corp Gate C Pre-tender
	Stage G Tender Documentation			
	Stage H Tender Action	3C Investment Decision		Corp Gate D Contract Award
Construction	Stage J Mobilisation		GRIP 6 Construction test & commissioning	
	Stage K Construction to practical completion		GRIP 7 Scheme Handback	
		4 Readiness for Service	GRIP 8 Project Close Out	Corp Gate E Project Close
Use	Stage L Use	5 Benefits Evaluation		

Figure 4.2   Typical workstage methodologies.

Interestingly these all consider preparation, design, preconstruction, construction and commissioning in readiness for use. They all recognise the importance of activities prior to the construction phase, but place different emphasis on the planning and development stages prior to Implementation. The OGC and NWR give greater consideration to the strategic need to establish a business case and output definition, and thus to defining what functionality and performance will represent success.

There are differences prior to the construction phase in the sequencing of option selection, procurement, design and tender processes and also whether or not the stages represent 'Go / No-go' points at which the project could be aborted, or just progress-control points. There is consistency across the methodologies that construction is seen as implementation but it is less clear whether design and preconstruction are seen also as a part of the implementation stage or whether they form part of the development stage. Interestingly not all of the methodologies consider use, as some see the completion and handover of the project as the end result whilst others see the operational stage as important.

Clients should therefore consider their own particular objectives and determine how they want to prioritise, depending on the degree to which they have pre-established approaches for funding and management of their projects. This assessment of the project timeline and the emphasis to be placed on the planning and development stages will also be influenced by the degree to which the client's procurement strategy sets design as an up-front client-led activity or as a subset of a contractor-led design-and-build approach.

There is a tendency for clients to believe that time spent in the early stages before commitment to construction is either less important or is delaying really getting on with the project. However, evidence suggests that this pre-construction period is the most valuable in establishing the project out-turn; defining what the client really wants, what he can afford and with what risk. Whether a project is likely to be a success or failure is often largely determined by the activities prior to commencement of construction.

This is reflected in the adage:

Plan long to build fast.

It should also be recognised that the expenditure profile is at its peak during the construction phase. Therefore a delay to review and clarify objectives is much less costly to the client if during the planning and development stages than during the implementation stage.

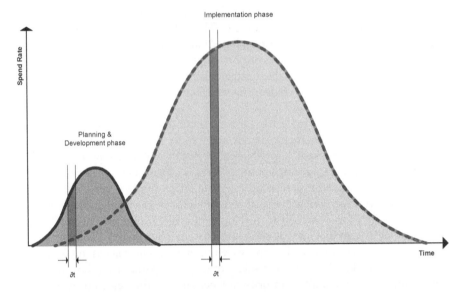

Figure 4.3   The relative cost of delay at different project stages – for the same time delay the cost consequences are significantly greater during the implementation phase.

Achieving project implementation

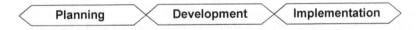

Figure 4.4   Planning and Development leading to Implementation.

An implementation stage is needed to deliver the project to reach project completion, so there is a tendency for considerable effort to be focused on this stage. However to get to the implementation stage there is a need for the client organisation to establish a framework for both the investment cycle and management cycle associated with their project.

This leads to recognising the need to include stages prior to the implementation stage.

Planning and development of strategies for the project are critical to enable successful implementation. Consideration is needed by the client organisation at the strategic level to establish the basis for managing the project before the project team is appointed. Those involved in implementing the project, the client's project sponsor and the project team, need to be able to operate within a pre-established clear management and investment framework.

An absence of a strong strategic management basis is a common reason for projects not having the right sense of direction and for there to be ambiguity in the objectives and in the control and decision-making processes.

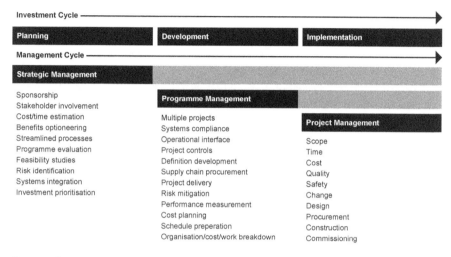

Figure 4.5   Investment and management cycles for project implementation.

Appointment of project participants

Where clients are unable to undertake in-house all roles needed for the delivery of a project, they need to assign and mobilise staff or appoint organisations who are specialists in order to execute the project. The timing of these appointments is a crucial factor in the client's overall management approach to the project and what skills, services and contractual commitments he wants to buy and when. It should therefore be carefully aligned with the client's proposed procurement strategy.

It is likely that some of these inputs will be needed as part of the planning stage, such as funding and legal advisors. It is also likely that the client organisation will need to take the strategic decision of identifying and assigning the right individual to fulfil the important role of client sponsor.

During the development stage it is likely that the client's project team will be assembled and key decisions taken on the procurement route for the project. This will set the context for the optimum sequence and timing for the appointment of other organisations and whether they are to be employed directly by the client or through other organisations. The development stage will also normally establish the required approaches for the control of the project and how important responsibilities for health and safety, planning and environmental management, risk, and media and public relations are to be assigned.

Operation and decommissioning stages

Figure 4.6   Operation followed eventually by Decommissioning.

There is a tendency for projects to be focused on achieving project completion, at the end of the implementation stage.

Expressions like '*delivered on time and on budget*' are used to define successful projects on reaching project completion at the end of the implementation stage. Many of the workstage models are set up to define only the phases and activities needed to reach this point.

But it leaves the interesting question:

Did the project provide the performance in use that was expected?

The best clients recognise that the success of a project is not defined by its implementation, but by whether it meets and delivers the required out-turn functionality and performance, within an overall investment cycle that considers not only capital expenditure but also operational expenditure. So there is an important fourth stage to cover – operation. This may be recognised as the concept of '*use*', but often there is a tendency in workstage methodologies for the emphasis to be purely on getting the project into use rather than on whether it goes on to deliver against the business case.

The increasing importance that society is rightly placing on legacy and sustainability issues is adding a further strategic consideration for the best clients. At the end of a project's useful life it needs to be possible to effectively decommission the project, such that there is no long-term negative legacy imposed on future generations. To be able to responsibly satisfy this there is a decommissioning stage that needs to be anticipated and planned for.

5 Representative case studies
Why the rest of this book is worth reading

Introduction

This chapter is not an attempt to provide detailed evaluations and analyses of case studies. Rather it is an illustrative selection of project examples. It uses information that is not new and is already in the public domain, selected to help the reader to gain an understanding of how major projects can go well or can get significantly into difficulty. The examples used indicate that this is not a function of whether there is a commitment to project systems, procedures and administrative support, nor hard work and effort, or even ultimately a preparedness to spend money.

Isolated mistakes are also not often the cause of things going wrong. Errors are made on almost all projects. The good ones have the ability and culture to compensate and work around them to recover. The bad projects somehow have a culture that compounds the difficulties – everything seems to just get harder.

Often there is a window of opportunity to get the clarity of thinking needed. If this is missed or bypassed in the belief that it can be picked up on the hoof, then things can cascade. Confusion followed by blame can become the project norm.

It was difficult to pick examples as few projects have solely 'positive attributes' and few are completely dominated by 'tough challenges' that might have been avoided. Almost all of the examples used could have been rewritten to reflect aspects of the alternative category. However, the aim has been to use project examples that can be broadly grouped under two categories – those that were delivered to budget and those that ran significantly over and became publicly associated with difficulty in delivery.

It is probably also making a few observations related to the theme

The person who never made a mistake never made anything.

It takes good people to be able to cope with the exceptional rather than the norm.

Being creative or accepting the challenge of delivering something complex and new is about being out at the edge.

However, an 'A-team' on a complex project can be less risky than a 'B-team' on a more average project, because they can handle, and thrive on, the challenge. Accepting that nothing goes entirely to plan reinforces the need to have leadership and a project team with the commitment to see through difficulties and get to a result.

Sometimes there is a great temptation to have a Plan B as a back-up option. However, this can be a signal of poor or uncommitted leadership. Generally, it is much better to get Plan A right with the right pre-planning and thinking, saying 'yes' to workarounds, 'no' to Plan B, as it just creates ambiguity and provides excuses for not being focused.

It is clearly difficult to pick example references to illustrate good and bad aspects without a very detailed knowledge of each, which the author cannot claim to have. Therefore the observations below should be considered as perceptions for illustrative purposes and are not purporting to be facts. All projects are a mix of good and bad.

Positive attributes

Given the above disclaimer, the author considers the following projects to be useful in helping to illustrate positive attributes to deliver success.

Example 1: CTRL – the 'Channel Tunnel Rail Link' project (subsequently known as High Speed 1)

The Channel Tunnel Rail Link (CTRL) project was the construction project to create and build the high-speed rail link between central London and the pre-existing Channel Tunnel rail connection to France. This railway is now known as HS1 to distinguish it from the planned HS2 between London and Birmingham. It was 108 km of new twin-track railway built to high-speed specifications previously unknown in the UK. Significant lengths of the route were in tunnel, including a section crossing under the Thames Estuary, and it had a number of major bridge structures. One of which, the Medway Viaduct, was, when built, the world's longest high-speed rail clear span. The London end terminated in an impressive regeneration of the famous St Pancras station. It has intermediate stations at Ashford, Ebbsfleet in North Kent and Stratford in East London. The route was selected to enter London north of the Thames Estuary in order to enable future connections to the north.

This extremely complex and varied project was the first new mainline railway constructed in the UK in over a century. The UK's rail industry had been remarkably good at just keeping an aging and under-invested infrastructure going – 'held together with sticking plaster and rubber bands' – but had no current experience of new major project delivery.

The approach taken in setting up and leading the project resulted in it providing the catalyst for significant urban regeneration (Kings Cross Central, Stratford City in East London and plans for Ebbsfleet in North Kent). It also provided UK construction with a new confidence that major infrastructure can be successfully delivered.

Aspects of the project that led to success included:

- It was fundamentally a worthy project on many counts
  - the completed Channel Tunnel desperately needed properly connecting to London.
  - the right route selection provided a major catalyst for urban regeneration that could help alleviate London's need for new infrastructure to support anticipated population growth.
  - the railway could provide new domestic commuter services and also release extra capacity into the overburdened existing rail network.
- The project strongly benefitted from powerful, if unlikely patronage (Michael Heseltine and John Prescott) so was depoliticised.
- It elected consciously to make good creative use of innovative but already proven technology, rather than attempt to develop new technology 'on the hoof'.
- The project was creatively structured financially through a PPP (public-private partnership) so that the SPV (special purpose vehicle) organisation in the client role could derive subsequent benefit from associated land development, which helped them secure cost-effective financing. This also motivated them to deliver the core railway project. The government's public interests were looked after at a high level through a small government representative team that understood their overview remit and did not try to meddle in the detail of the client role.
- The client SPV consortium understood and defined their primary objectives and established at the outset a strong vision and inclusive project culture to support these.
- The client SPV consortium invested in appointing an extremely experienced project development and management team able to be innovative. This team properly set out and then resourced the effort needed for planning and environmental consenting and the subsequent project delivery, with the designers and management brought on-board from the outset.
- Throughout the project, effort was placed on identifying and quantifying risk and then placed with those most able to manage it. This included not entering into construction contracts until the associated planning and consenting obligations were fully established and could be fairly defined.
- The project leadership understood the need to invest up front in a strong project 'infrastructure'. By doing so it took away excuses for non-delivery. (E.g. the project invested in establishing a common information and data platform as the largest ever set up for a UK construction project, investing in the latest proven IT technologies and document-control systems; it invested in building a temporary project office to bring together 900 staff under one roof; it invested in a temporary factory as the largest manufacturing facility of precast concrete tunnel segments and dismantled it afterwards; it invested in

constructing a railhead facility for just-in-time rail systems delivery and then returned it to landscaped countryside after construction.)

The project was not without major challenges, including being split into two sections with different financing arrangements, but the approach and leadership was powerful enough to achieve confidence for a consistent delivery of both. The two sections were each completed on time and to budget. Many of those involved have gone on to fulfil major leadership roles on subsequent large infrastructure programmes.

Example 2: London 2012 Olympic Park

The London Olympics in 2012 was internationally recognised as a success. This was a major triumph, particularly since there was much effort by the press and media early on to cast doubt and to pick on any negative angles they could. Once the project was complete, this changed. Interestingly the contribution of the construction success was recognised in the opening ceremony, both by praising the UK's heritage in great construction engineering (Brunel) and also by the presence of representatives of the construction labour force.

The Olympics were mainly located in a new park (now known as Queen Elizabeth Olympic Park) in East London, adjacent to the newly regenerated Stratford City. The chosen site was an extensive but fragmented area of dereliction, pollution and poverty. This had resulted from a history of run-down industry and under investment, despite it being relatively close to the centres of London business. A legacy from the Olympic Games would be to transform the site and give it a real value to London.

This ambition helped London win the bid for the Olympics. At the same time, it made the task of delivery even more challenging. It could have been the sort of project that was seen as a 'hot potato' and one to avoid. It is to the considerable credit of those who were in senior leadership roles that it developed a culture and dynamism to be the exact opposite.

Aspects of the project that contributed to its success included:

- The project benefitted from the absolute need and clear application to achieve the deadline. There was no ambiguity over when the project had to be delivered. This imposed discipline was a good constructive focus to the decision-makers.
- Its massive complexity was, importantly, broken into three constituent parts, which can be portrayed as 'the site, the theatre and the play'. A good scoping discipline was achieved between the LDA (responsible for assembling the site), the ODA (responsible for construction of the facilities) and the LOC (responsible for putting on and running the Games). Each had strong motivational leadership responsible for their respective parts but joined by a common goal, so willing to be helpful rather than antagonistic at the

interfaces. This has subsequently continued through a fourth constituent part with the formation of the LLDC (responsible for the site's future and legacy).

- In assembling the site, it was recognised that it extended across a number of London boroughs and consenting authorities. This issue was not ducked, and up front a unified approach by all was agreed and achieved.
- Site access and control was invested in early and placed under one responsibility. The associated requirements were clearly spelled out in advance to all involved in construction activities, so everyone knew the rules and there were no excuses.
- The construction delivery client assembled a single integrated team (CLM) as a consortium of construction planning and management expertise. They then appointed top-quality industry-leading firms to deliver in both the design and the construction. They used contracts that were fair and transparent, such that they could have flexibility through controlled processes to make adjustments to pick up minor revisions for unforeseen items and thus maintain delivery momentum.
- The project generally did not seek glory early. The project delivery team put little effort into countering the press detractors and focussed on delivering rather than making early promises. It left taking the credit until after success was achieved.

Example 3: Heathrow Terminal 5

Heathrow airport is London's major transport interchange and needs no introduction. It is one of the world's flagship aviation destinations and rights to landing slots are highly sought after by the major airlines. The airport's passenger-handling capacity needed to be improved and additional stands made available for aircraft. Terminal 5 was a means to achieve this, located at the west of the airport with direct access from the M25 motorway and new underground rail connections all part of the project scope. It was to be the home for the UK's signature airline, British Airways. The business case was both pressing and clear.

Planning and consenting of this project was an extremely protracted process, so many thought it was a lost cause. It would have been easy for the scheme to have become labelled as the project that never would be and therefore one to avoid. However, despite that, the strength of leadership and unwavering belief in the project's need and purpose has resulted in it now being acclaimed as one of the very best airport terminals in the world.

Some of the aspects of the project that help it merit this are:

- Whilst the planning and consenting process was lengthy, a discipline was maintained to not embark on significant construction until consenting was adequately in place.
- Despite the lengthy consenting, the client saw the project as central to its business and therefore invested in it as a primary commitment. When implementation was embarked upon, well-developed plans were ready and

in place, so that design and construction could be centrally managed by an integrated client-led team with a clear ambition for modularisation and off-site pre-assembly to simplify and de-risk site activities.

- Quality was seen as an important outcome and one to reinforce the client's international reputation and that of British Airways as the principal user when competing in the world-wide aviation market. So top industry-leading firms were appointed for the design and construction.
- However, the client set up the means and in-house expertise to be able to coordinate and control these appointed firms. The designers and the contractors were separately appointed but were kept close enough through the client-led team to work together with the common objective to benefit the out-turn result.
- The project had to compete for labour with other projects in central London. The need to get a major construction labour force committed to working outside of central London was recognised early and special measures were put in place to encourage and attract the required resources.

Initial baggage-handling problems were an embarrassing hic-up, due to a failure to recognise that operatives busy working in the existing facilities could not be available for training in the use of the new facility. However, the inherent competence of the project and the completed facility meant that this was quickly overcome and the project was acclaimed as a national success.

Example 4: Millennium Bridge, London

This is a project that might be considered unsuccessful as it had major problems on opening. It is the iconic new pedestrian bridge crossing of the Thames in central London, between St Paul's Cathedral and the Tate Modern. It is a suspension bridge like no other, with the cables having a very low profile on outriggers almost within the depth of the bridge deck.

- The project has a strong purpose in improving pedestrian connectivity between the City of London and the South Bank. It has contributed to the resurgence of activities and amenities along the south side of the Thames.
- It was extremely innovative and at the boundary of established codes and standards. Dynamic lateral vibrations occurred, compromising in-use performance, which were identified almost immediately on opening. When this major problem arose, the client, stakeholders and key team members quickly worked together and collectively supported the effort needed to resolve it in a 'no blame' culture to get to the earliest possible solution.

The consequence of this strong leadership culture was that the project headed towards finding a solution and not the law courts. In the end, all of those involved were the beneficiaries of the approach taken and can be proud of their involvement in an exceptional iconic project for London. It has since enabled international bridge codes to be developed further and has therefore also contributed to global industry expertise and knowledge.

Example 5: Glaxo Research Campus, Stevenage

This project was undertaken back in the early 1990s. It is not high profile but at that time it was the UK's single largest building construction project. It was an extremely ambitious project to develop a site located just south of Stevenage, midway between London and Cambridge, to provide Glaxo Pharmaceutical (now GlaxoSmithKline) with a major research and drug-development campus. Aimed at keeping the company at the forefront of the internationally competitive pharmaceutical industry, it brought together all of the company's research faculties onto a single campus – the objective being to maximise cross-fertilisation between these research departments and their different sciences.

The company recognised the scale of its ambition and that it had not done anything comparable previously in the UK. It elected to review experience from other industry sectors and invited senior figures associated with these to attend advisory, knowledge-sharing workshops. By doing so it determined a novel approach for the project. The resulting project set new standards for successful delivery in the construction industry. These were largely initiated by the client.

Some of the factors influencing this success can be summarised:

- The client brought in senior people to form the project leadership from another industry (oil and gas) who had proven experience in successfully delivering difficult projects. This leadership had succeeded by incentivising those others involved in these projects through a predefined sharing of the upside benefits and being motivated to avoid the downside. Everyone had 'skin in the game'. This culture for shared success was established as core to the Stevenage project.
- The client's leadership selected the project team on the basis of proven construction industry capability and 'can-do' intellect. They wanted something new, able to match their aspirations, so not the traditional incumbents with an immediate track record in their existing business sector.
- The client then invested in team building to bring everyone involved up to speed. To understand what was at the time quite a radical departure from the norms and to create a project culture with an inclusive leadership all having a common understanding of the objectives.
- The organisational structure of the team related to projects, not portfolios. Leadership was structured not into titled directorates but with a direct alignment to the component elements of the project. It was therefore delivery focused and not focused on functional importance.
- Considerable effort was put into clearly defining the project scope and performance criteria. The client had successfully completed a research facility in the USA. It therefore engaged the American team for the early project definition.
- But knowing that the American consultants would not travel and would not

have the UK experience to follow through, a clever approach was used for procurement and transfer through the 'gate' principle. An integrated multidisciplinary UK design team was competitively appointed, based on preliminary information from the American team. All of the UK competitors bid against a common set of documents, even though the client knew these were not final. This enabled the client to select the preferred UK team. They did this on the basis of both pricing and understanding of the client's objectives.

- Senior members from the appointed UK design team were required to start out in America alongside the American team as they completed the project-definition stage. The UK team's involvement being to understand and sign up to the adequacy of the 'gate' review. Outstanding actions from the review were assigned either to the Americans to quickly resolve or formally placed with the incoming UK team. This eliminated excuses for ambiguity or uncertainty in the next design-development stage as the UK team were there and involved.

The project was a strong subsequent influence on many forward-looking clients and organisations in the UK construction sector. More recent and widely adopted new forms of contract, such as the ICE NEC reflect many of the positive aspects of the Glaxo Stevenage project.

Tough challenges

It is obviously potentially contentious to single out projects under the category of 'tough challenges' that might have been avoided. The selection below is therefore more to illustrate aspects that would have been better avoided than to attempt to absolutely classify these projects as flawed. As high-profile projects, they arguably rank highly on many other counts.

It is interesting to note that the aspects cited in many cases stemmed from confused objectives and expectations. It might have been seen from the outset that these were likely to militate against good out-turns.

Example 1: West Coast Mainline upgrade

The objective of this project was to improve performance of one of the busiest UK mainline intercity railways. To achieve this whilst also maintaining the current train services was always going to be difficult. It was analogous to trying to convert A-roads into grade-separated motorways on the same alignment, while trying to keep the traffic flowing. And, if this was not challenging enough, it then proposed to introduce a new technology of moving-block signalling.

So, how to make a project difficult:

- Define a scope that in effect requires '*Holding oneself up by one's own bootstraps*'. Having a project that involves so much interface with other operational systems that every aspect is doubly complicated and may not even be achievable when looking at the detail level. The team may find itself in a spider's web of constraints.

- Expect operational public services to co-exist geographically with major construction sites, and by doing so also make it a logistical nightmare to bring in materials and major construction equipment.
- Expect a system of railway possessions with handover/handback arrangements geared mainly for short-term routine maintenance to be used for major new construction. Eventually it was realised that blockades taking out of operation sections of railway for weeks and months were needed.
- Applying an unproven new technology 'on the run' as an integral part of the project. Moving-block signalling technology is a new system that in effect attaches the protected zone to the train, whereas traditional signalling is based on the zones being fixed-block lengths of the railway and controlling trains going in and out of them. The new moving-block concept will enable greater numbers of train paths, with trains running closer to one another based on greater intelligence of their speeds and characteristics. It sounds simple in theory but represents a fundamental switch in the relationship between the railway and the trains operating on it. To introduce this on the West Coast Mainline Upgrade would have been attractive as it would have helped increase the line capacity. But it proved too difficult and was eventually removed from the project scope, only after considerable cost and effort had been expended to try to include it. To introduce new technology while at the same time attempting to maintain existing train services using current fixed-block signalling had not been thought out in advance in the client's project scoping sufficiently for it to be deliverable.

The project was subject to overruns, delays and costs exceeding the original budget. In many ways this might be considered more typical of major infra-structure projects. Confirming the broadly held perception that almost all are in some way mismanaged or more complex than expected, so have to cost more than was planned and are late. The West Coast Mainline Upgrade project was de-scoped and rationalised and the final result achieved was a much-needed improvement and benefit to one of the UK's most critical railways. It was a credit to those involved that they were able to see it through to a successful result. In hindsight the final scope was probably the sensible answer. Earlier assessment of the project interfaces, risk and scope could have made getting to this out-turn more straightforward.

Example 2: Channel Tunnel connecting UK to France

This project was seen as a key opportunity to have the UK more clearly linked as a part of Europe and the European Union. It was also an opportunity to prove that modern-day construction could succeed where previous attempts had failed. However, a laudable project was subject to confused ambitions from the outset. Road or rail, bridge or tunnel or islands were all considered and placed in competition, so where was real responsibility for the scope? Surely an informed and competent client organisation might have worked out which of these it wanted. Instead a lot of money was spent by a range of ventures all trying to position their quite different schemes against one another.

There was high-level political ambition for the project, but this was motivated by contrasting rather than aligned objectives on the two sides. The consequence being that the client body was an amalgamation of UK and French organisations responding to the agendas of their respective political masters.

How to create ambiguity in scope and by encouraging lobbying from contrasting ventures:

- Encourage competing schemes that are developed on fundamentally different proposals and functional objectives, with different financial motives – so a situation of '*comparing apples and oranges*' at just about every level.
- Try to encourage creative financing, but get ambiguity and possibly confused relationships between the appointed client side and construction side. Some of the same players, initially at least, were involved with both the client organisation (Eurotunnel) and the construction contracting consortium (TransManche Link), so the motivation to control costs was not always clear.
- Help finance the scheme with shares sold to a general public (often French private citizens) who progressively saw their shares become worthless. The completed project might be viable, but only if the construction cost was written off, resulting in no return for many of those who had invested in its delivery. Not a particularly good business model.
- The tunnel construction was technically excellent and the accuracy of meeting midway was impressive, even though there was some challenging rivalry between the UK and French sides of TML.
- The completed project has not been without operational performance challenges, arguably associated with a flawed interface between the civil engineering and the transport systems. The fire scenario for the Eurotunnel Shuttle train operation has not been compatible with the fire rating and resistance of the tunnel construction. This has also compromised the future capacity of high-speed passenger train operations as the current duplex (double decker) trains are of aluminium construction so not with a fire rating for the tunnel.

So a project that in many ways has been a fantastic feat of engineering and a major achievement, both politically and technically, has not quite achieved all it set out to.

Example 3: British Navy Astute Class Submarine Programme

This project is typical of many major capital investment programmes. An improved performance was needed, with a slightly larger submarine than the current class. However, since budgets were limited, the argument used was that by just elongating an existing design it would be possible to be confident regarding delivery and keep costs to a minimum. This all seemed logical and the go-ahead was given, with the promised costs included in future spending.

How to let go of a simple plan:

- What started as a simple increase in length of an existing design got confused by creep and extension to the project objectives. There was a tendency for the syndrome of, *'We know that we only have a minimal budget, but while we're at it, it would be foolish not to just include "xyz", which in itself will not be a significant extra, or we can justify it through some other budget.'* But having done this a few times, the extras began to make the original idea less relevant. It was no longer just a physical extension of the existing design. There were some radical changes being introduced that though superficially not major were undermining aspects of the original design so it was no longer the optimum solution.
- The supply chain was not sufficiently joined up to enable a coordinated design. The design many thought they had committed to was the original, simply extended by a few metres. But the inclusion of some 'extras' meant that some of the suppliers were designing to these while others were not. The knock-on consequences of the extras needed to be co-ordinated across the whole supply chain. A perfect excuse for change, meaning claims for time and money, and a poor negotiating position.
- To compound this a new 3D CAD system was introduced to try to set up a common information platform. This was meant to resolve technical interfaces but these benefits proved more difficult to achieve than either the MOD or the contractor envisaged, and further added to complications with the supply chain.
- There were also difficulties with the capabilities and capacity of the systems contractor, with the contract having been awarded without an adequate appreciation of these limitations.

What was proclaimed at the outset as a confident programme proved to be inadequately planned and in consequence naively optimistic.

Example 4: Jubilee Line Extension, London Underground

This project was the significant extension of a key part of the London Underground system, both south of the Thames and to East London. Critically, it was important to provide much-needed improvement in access to Canary Wharf, a major new business centre for London. It also helped develop residential, retail and cultural regeneration of the former London docklands. Since opening it has been a fantastic success.

However, its delivery was not straightforward:

- The project embarked upon a feast of design originality that seemed creative and exciting. However, the reality was uncertainty over whether it was setting out to be an architectural showcase. There were different architectural teams and ambitions for each station, but were the cost consequences and repeated learning curves associated with this really understood? Were the available London Underground design standards adequately defined to be able to cater for this multiple and varied interpretation?

- The project set out also to adopt a new train control technology of 'moving block' signalling. Like the West Coast Mainline project, this was eventually abandoned to enable delivery of the project.
- The project's original leadership was constrained by budget and thus not empowered to take major decisions. It was then criticised for failure. Ironically the replacement leadership was given almost a 'blank cheque' and then hailed for achieving what the original leadership had not.

Complicated by overly ambitious technical and architectural objectives, the project needed a lot of money injected to force it to completion.

Example 5: New Scottish Parliament, Edinburgh

This is an iconic project. However, it became evident that the ambitions for unique exceptional architecture and the available budget might be in conflict.

- So it could be questioned whether winning the high-profile international design competition gave the architect a mandate to produce his competition scheme, even if it was more than could be afforded? Was the client able to recognise that the selected winning architect's proposals would radically exceed the available budget as the full scope and details of the project emerged? It is interesting to consider if by selecting a winning scheme a client in effect takes responsibility for being able to afford it, and for the ensuing complications if it cannot be afforded.
- Uncertainty over the desirability of the unusual design was a public reaction to the scheme as it evolved. But the architect had been selected and was expecting to be able to produce a characteristic design with its associated potentially controversial 'trademark' features.

This is not an uncommon situation. To appoint a design team on the basis of attractively exciting proposals that were likely to be more than could be afforded. Then the subsequent debate over where responsibilities were for containing cost and accepting the more radical aspects of the design.

Example 6: Wembley Stadium, London

Modernisation of the national stadium was needed. But this then led to various aspirations as to what it should be. The previous stadium had a rich and varied history which gave the project an importance, but the challenge of *'An importance for what?'*

- Early uncertainty over whether it was to be for football only or also for an athletics running track led to competing 'client' demands. Consideration was given to it being an Olympic Stadium but this would have compromised it from being optimal as a football stadium, since the sports-field geometries of these are different.
- Somehow this lack of clear objective direction to the project appeared to continue into construction. Responsibilities for uncertainties and evident cost overruns were contested. The various sides were at odds and problems were allowed to polarise and fester. The consequence being that they became amplified, rather than being positively addressed through collective effort and contained.

The football stadium ambition won out, as did the technical sophistication of a moving roof to prevent events from being spoiled by poor weather. But the resulting project was certainly not cheap. However, the football stadium decision did result in a clear position for the 2012 London Olympics, with no ambiguity over the need for the main Olympic athletics stadium being included within the Olympic Park at Stratford, East London.

Example 7: Schoenefeld Berlin Brandenburg Airport, Germany

Reunification of Germany re-established Berlin as the country's seat of political power. Following the Second World War the city had been split into sectors and in consequence had developed three airports, none of which were adequate to service the future needs of a unified capital city. The decision was taken to rebuild and enlarge the airport at Schoenefeld, the only one of the three with the physical space to expand, as the new Berlin Brandenburg Airport.

The opening of the new airport in Autumn 2012 got to within just a very few weeks when, to huge embarrassment, it was realised that it was not ready.

- This situation was particularly severe as the opening ceremony was all lined up and all of the many organisations, retailers, and so on planning to operate within the airport had recruited, employed and initiated new staff. The opening has been substantially delayed, and is now planned for 2016, so this was not a minor hiccup.
- Planning for construction was one thing, but on such a complex project there is also a need for planning for operational acceptance. Much of the delay has been associated with the ability to comply with fire alarm/safety issues. Planning for and populating the 'safety case' for operational approval is a major activity and needs to be considered from the outset.

For major projects such as Schoenefeld the operational safety case is not a 'bolt on' extra. It is crucial to plan for taking along the various authorities responsible for issuing certificates for operational approval. Early commitment is needed by the client to establish a dialogue between the operating team, the external authorities and the construction team.

Summary thoughts

It is interesting to consider the degree to which the early thinking by the client was relevant to these example projects.

Going into projects early may seem like a show of confidence by clients. However, this can be false confidence if pushing towards the delivery phase when not enough preparatory work has been completed within a development phase. It is always crucial to be clear on the objectives for the projects – the '*why*' and the '*what*'.

Setting out with ambiguity and uncertainty must be a client responsibility. The same for setting out on a course with a poor understanding and alignment of risk that will almost inevitably lead to complications.

6 Project definition

What is wanted and when – being able to define it

Introduction

> *How can you expect someone else to achieve something for you if you cannot in some way define what it is you are wanting?*

Many projects do not succeed as a consequence of the client embarking on the project without having rigorously considered this, and describing his strategic intentions unambiguously.

Successful construction projects have a clear vision and delivery strategy from the outset.

Clients need to identify the business needs that any proposed project will fulfil and determine how that project sits within their overall business strategy. A clear vision for the project, which states the objectives and outcomes, is critical in justifying the project in terms of investment and building the business case.

An adequate definition of what is wanted from the project can be encapsulated by a combination of the following:

- Vision statement
- Business case
- Client brief

Vision statement

Does the client have a vision of what the project is about?

It may be useful for the client to prepare a 'vision statement' to test and define

- What will represent success?
- How real is the project and is there client support for it at the right level of authority?
- How 'mission critical' to the client's business is the project?
- What values does the project want to embrace?

Normally the 'vision statement' is an aspirational document with the aim of establishing the high-level drivers behind the project. A good vision statement may also be inspirational; capturing the imagination and exciting people to want to take up the challenge and motivating a desire to be involved. The vision statement should help set the tone and intellectual level of the project. It may set values in terms of social responsibility, characteristics of behaviour of those involved, the context of the project in relation to broader cultural, environmental or historical factors and whether it is intended as a legacy.

The vision statement should be able to stand the test of time, so is normally prepared early and not routinely updated. It can then be a useful check that the project has not been derailed or 'lost the plot' through scope creep or interference as the project proceeds.

Projects that are unsuccessful are often ones where the final product has attempted to satisfy objectives that are barely recognisable when compared against those of the original remit.

Business case

The client needs to understand the fundamentals of his business case for the project, and the sensitivity of each parameter to a successful outcome.

At the most basic level – *'How real is the project?'* Most projects need to be affordable and to provide benefit that is in some way proportionate to the investment. At the inception of the project does it pass this test, and how constrained or borderline is this equation?

The client should be frank about the robustness of the business case as it will help his custodianship of the project and decision-making as it proceeds. An important aspect of this is to invest time and expertise at the inception of the project to set a realistic cost budget. Many projects are always on the 'back foot' because the original budget was inadequate; the team's efforts are then constantly deflected into explaining and defending a cost difference between reality and a naively optimistic budget. Vision, budget and scope must be at one with each other. High aspiration coupled with an inadequate budget is not compatible, and will almost certainly result in an aborted project or at best one finished with an air of failure.

In this regard it is essential that clients act responsibly and do not expect other parties to commit time, valuable resources and intellectual investment to something that has no basis. Good clients are mindful of this and do not abuse the goodwill of others when just 'flying a kite'.

The client needs a business case that aligns the project with his key business needs and operational needs. It will help establish the timeframe, funding and revenue streams that justify the project.

The business case needs to be a principal control document of the client for the project. The project will need to be reviewed routinely throughout its development to demonstrate continued compliance with the business case. The objectives and outcomes should be tested as the project progresses, to check that the project is still on track and that assumptions made in the business case remain valid. Many clients use gateway reviews to check the status of their projects at different stages and to provide the opportunity to verify and validate that the project still meets the requirements of the business case. These reviews should be used to ensure that the strategic vision and objectives are still being followed and are still valid.

Client brief

The client brief should define what the client needs the project to achieve in order to deliver the success and values identified in the vision statement. The client brief should also be anchored by the business case and not set objectives that are clearly beyond the envelope justified by the business model.

There can be different approaches in formulating the client brief.

In some cases the client may have business-related or functional performance-related objectives and might be quite open to how these are realised. It might even be seen as a positive advantage to leave as much open as possible to allow creative solutions to be developed from those with skills or experience that the client does not possess. In such cases a 'Performance Specification' approach may be taken.

In other cases the client may have a clear picture of how he sees the project manifesting itself and the particular design configurations that are required. He may already have done similar projects and want many of the features of these incorporated as they are known to work. Then it may be more appropriate for a 'detailed brief' to be prepared.

Like the business case, the client brief is a principal control document for the project. Clients need to consider and understand what level of control they want over the manifestation of the project and prepare a brief which accords with this. Basically, if you know exactly what you want then say so and don't waste everybody's time trying to guess. But then accept that if you have been absolutely prescriptive you cannot expect lateral thinking and creativity; or vice versa. So don't follow one approach and then expect the other, and be disappointed with the ensuing project.

However, the client brief can be an evolving process if properly controlled. If the client recognises that he does not know at the outset how to fully define his requirements then he can seek expert advice and input. The project can go through progressive levels of definition:

- SOC (statement of criteria) – these may be largely performance-based rather than defining solutions
- BOD (basis of design) – this may be a representation of a solution that takes the SOCs forward, tests the ability to mutually satisfy them and provides an illustrative design

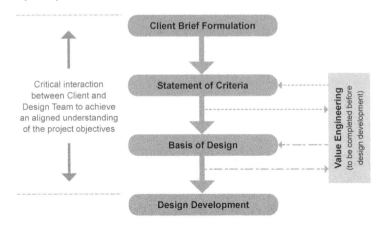

Figure 6.1   Importance of early project definition – making the most of the
pre-construction development phase to be clear on the design objectives.

The BOD may be developed on behalf of the client and through close consultation, including the preparation of options or alternative designs, so that the
client is directly involved in the selection process and the incorporation of this
work into a more detailed client brief. In doing so it is important to check and
maintain the goals of the vision statement and to not lose track of the limits of
the business case.

A key aspect of the client brief is to encapsulate the requirements for budget
and best value, so these can be communicated to the project team. Clients
should ensure that the budget derived from the business case is realistic and
delivers best value. Clients should consider the need to use 'whole-life costing',
which involves considering design, construction, operation, maintenance and
decommissioning costs from the outset. Sustainability and health-and-safety
aspects should also be considered for the project as a whole.

Clients should also set out a realistic programme in the client brief, taking
due account of the revenue-stream assumptions behind the business case. The
client should understand how the accumulated outgoings will be funded ahead
of income generation from the completed project. The programme should, so
far as possible, identify all major interdependencies and be sufficiently loose
to be attractive to the marketplace and attract reasonable rates. A very tight
programme with little contingency means risk, and risk usually needs to be offset
through elevated prices, so there is little to be gained by making the programme
tighter than is really needed.

Important in the preparation of the client brief is to establish the project scope
at the outset and to then actively manage compliance to guard against scope
creep.

One cause of scope creep and late change is the late recognition of operational requirements. It is strongly advisable to have these clearly articulated as an integral part of the client brief. The requirements for future-proofing the project's performance in use should also be considered and defined. It is therefore appropriate to have early input from the envisaged operators of the completed project.

Value engineering

Value engineering is a tool which should be considered to ensure that the most economically advantageous methods and materials are used in the project and that the appropriate quality is being maintained. However, value engineering is a much misused and misunderstood concept. It is not a cost-cutting device, but rather a way to optimise the relationship between value to the client and cost. Value engineering exercises are therefore very dependent upon close client involvement and understanding of the full consequences of any proposed changes. It also needs to be recognised that change does not come without cost and potential delay whilst revised proposals are thought through and implemented. Therefore experience suggests that value engineering is best done early and before major commitments have been made contractually to organisations that would see advantage in change.

It can be a very clumsy tool for a client who does not know what he wants, to try to re-establish a basis of control. Post-applying value engineering to a design developed from a client brief that was wildly ambitious relative to the business case and the available budget is not a route to achieving a successful out-turn.

Previous experience

Previous experience can be a major factor in whether or not the client is able to articulate and define what he wants – '*Has the client done it before or is it a new venture?*'

Surprisingly often, major projects are undertaken by clients or teams that have not previously worked together and have no experience of the type of project. The new project may be needed because of long periods of under-investment or in response to a completely new market, so there is not a readily available track record. Or the project may be so large that it has to be undertaken in a unique way and so requires a new organisational structure with new terms of reference.

An SPV (special purpose vehicle) or joint venture may be set up with a remit to fulfil the client function. This can be useful when the project is unique and requires a new entity to be established, or when there is a requirement for the project to be kept separate, as an 'off balance sheet' activity. The client body is formed from recruiting in senior staff, resulting in them having individual expertise and experience but not necessarily a track record of having worked together. It may well be that individuals within this have fulfilled the client role previously, but this does not automatically mean that a new collection of individuals is sufficiently aligned as a team to know what they want with a united front. This is not unusual but should be recognised at the outset and early investment made to create a sense of unity and an effective client body before progressing far with the project.

In other cases the client has completed similar projects before. Good clients analyse these past projects to identify their successes and weaknesses and can articulate the differences they want to introduce. Such early feedback and its inclusion into the client brief can make a major contribution towards a successful project out-turn.

Another approach is for the client to invest time early, together with his project team, in researching similar projects (good and bad) and collectively taking and applying key lessons learnt from them.

Financial or legal

The client needs to understand the financial and legal constraints within which the project will have to operate, both during its implementation and during operation. The business case and client brief need to be compatible with these, so that what the client wants is permissible and achievable.

The client has an absolute obligation to have the necessary funds in place. Asking others to carry out services and incur costs when the client knows he has not yet got the funding structure in place to pay them is unacceptable. The business case needs to reflect this and define the relationship between expenditure and income over time.

Characteristics of major projects

In defining a project, it is worth considering whether it fits into the normal spectrum or whether it should be classified as a major project requiring a different approach. If one or more of the following apply then serious consideration needs to be given up front to a different, more sophisticated approach to the project and its planning and procurement:

- *It may be a 'one-off' for the client, or involve a specific commercial framework (such as an SPV) to be established outside the client's core business:*
  The client will probably need to form partnerships with organisations which have experience in such major projects, but leave commercial commitment open until the scope and the risks are defined, in order to retain some element of market competition.

- *The project requires special consent to achieve the necessary statutory powers for its implementation:*
  The lead-in to the project will need to allow for the time to gain approvals, including environmental impact and land acquisition (LOD = limits of deviation for the three-dimensional envelope within which the completed works will be contained; LLAA = limits of land available for access to the

completed project, such as for maintenance or emergency situations; LLAU = limits of land available for use in order to be able to construct the project).

- *The project is often large enough to influence and distort the local construction market:*
  It may require up-front investment in construction logistics, early approaches to major suppliers and even the setting up of just-in-time delivery using a remote consolidation centre. Early consideration is needed to develop a procurement strategy which aligns the work packaging with national and international contractors' skills and capacity to deliver.

- *The scale of a project usually leads to significant numbers of interfaces and/or stakeholders:*
  An integrated team is needed to put early effort into a strategy to identify interfaces and their associated risks that places the interfaces where they can best be managed. Common computer-aided design (CAD/BIM) and geographic information systems (GIS) will be needed for seamless control and to manage data and standards centrally to ensure consistency.

- *Not all of the objectives will be known at the outset, and they will vary over the life of the project:*
  This necessitates a controlled approach to defining criteria and the design basis, so that at any point the objectives are known and changes can be managed. It also entails a 'packaged approach', where contracts are awarded progressively once relevant objectives are decided.

  An interesting observation was made by someone involved in putting satellites up into orbit. When you press the launch button everything has to work, as you can't go and sort it out later. He explained that one of the best ways to ensure that everything works is to agree on what is wanted and don't keep changing it. In this way everyone knows what it is that must 100 per cent work when it gets into orbit.

  He compared the situation to buying a Christmas tree. First he was asked to buy a tree, so he did, but then his partner asked where were the tinsel and decorations, so he went out again and bought these. Then he was told that it needed lights. By now he was beginning to think ahead so before heading out again queried what sort, what colour and did they need to flash? A little project he thought would take an hour ended up taking all day and costing more than double the cost of the tree. Just as well he was not sending a Christmas tree into orbit!

So the challenge when defining the project requirements is to know whether you are asking for 'a Christmas tree' or for 'a fully functioning and decorated Christmas tree'.

7 Governance of the project team

How to set up and organise the project team

Introduction

The term 'governance' has been already referred to in the earlier chapter describing the client. In setting up the project team it is important to first be clear on the structure of the governance for the project.

The term 'client' is used throughout this book. However, the client structure may in itself be complicated and there may be a project sponsor role in addition. Large projects may have a board-level representative, a sponsoring authority or even a department of the government determining the need for the project. This is responsible for determining the means for the control and direction of the various aspects of the project and its strategic objectives.

Below this, complex projects may well need a shape and structure for their governance. Specialist expertise may be needed to lead and take responsibility without ambiguity, with organisational structures set up. These should consider and address the experience, leadership and quantum of effort needed for:

- defining and obtaining consents and approvals for the project
- stakeholder engagement
- getting the project built, delivering the project
- gaining compliance acceptance and operational readiness
- maintaining and operating the completed project
- running the primary revenue-generating activity
- managing the land and property assets associated with the project, land acquisition and creating additional related revenues

All of these are necessary contributions. Some organisations set these roles up as independent departments, each with their own authority and autonomy. However the challenge is to avoid internal 'power barons' if these create divisions in ambitions rather than collective focus. There is a need for the overall successful delivery of individual projects to be the singular combined ambition, rather than the internal battlefield platform.

Board of control

There is then a need to bring these different aspects together so they can perform as a collective team with a common purpose. This is normally achieved by a board of control for the project, with each of the leaders of the responsible organisations represented.

The chair of the board has responsibility for the balanced performance of the whole and the effective contributions from each. The board of control needs to provide the project integration to bring together these different inputs.

The sponsoring authority can hold the board of control to account and can ask to see the performance of each of the responsible organisations. This can provide the transparency needed to have confidence. A responsibility of the sponsoring authority is to set up the delineation between it and the board of control.

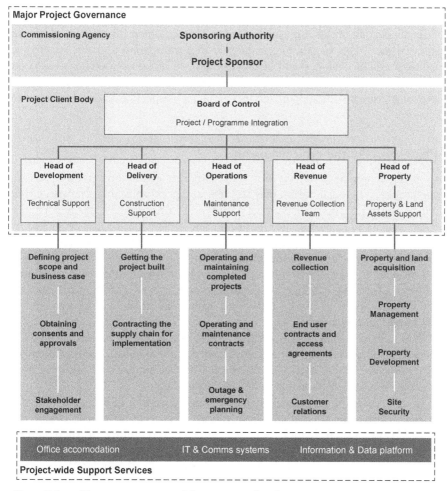

Figure 7.1   The governance portfolios associated with major projects.

The structure of governance needs to play to the strengths of top individuals in the industry who can be made available for the project. Some are experts in leading project inception and approvals, stakeholder engagement and public consultation processes. They may well enjoy coping with the uncertainty and ambiguity of the early stages. Whereas others want something that is predefined and are completer-finishers. They understand the single-mindedness needed to drive major construction organisations to deliver quality within time and cost.

Figure 7.1 on the previous page illustrates how this governance might be structured on really large projects which require a complex leadership structure that might progress from one emphasis to another through the life of the project.

The Head of Revenue role might be the input of a major user of the completed asset, such as a train operating company (TOC) paying for pathways, or an offshore wind farm operator paying the offshore transmission operation (OFTO) on the basis of HV connection availability. In the case of a regulated utility it might be the role of revenue collection from the paying consumers.

To support all of this it is usually good to have common information platforms so everyone is able to talk to one another in a transparent and efficient way. It is important to have a Board of Control able to make key decisions based on only one version of the truth when it comes to data and facts. The appropriate common project-wide support services are a good aspect of achieving this sense of team.

Phases to take the project forward

Complex major projects may not start with the 'client' organisation owning the project. The organisation may feel it is an opportunity that they can either create or attract into their business portfolio. To then achieve this may require them to work out how to bid or position themselves for it. Such a development opportunity may arise by pre-bid analysis of emerging market conditions or be driven by a significant government programme.

Once an opportunity is identified, the client organisation and appointed advisors need to be structured with controls and a governance that can manage the bid activity to gain the project and attract the financing arrangements to fund taking it forwards.

The next step, having gained the right to develop the project, is post-bid to set up the governance arrangements for the procurement and management of its delivery. These then lead on through to acceptance and takeover from construction and on to operation and maintenance.

In some cases the model may include a planned transfer in ownership in order to realise the value created and profit return. This profit return can enable the organisation to then finance their next targeted opportunity. So there can be different combinations of Finance, Build, Own, Operate, Transfer, all needing structures for governance of the project team with the appropriate lead within the board of control.

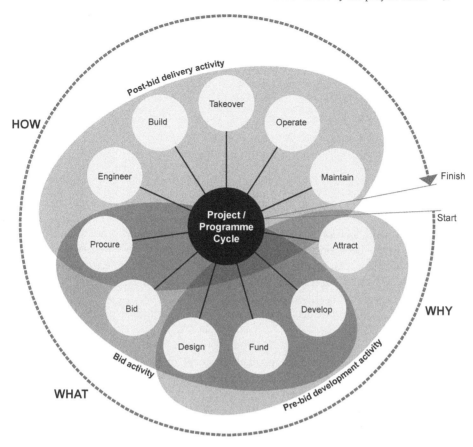

Figure 7.2   The project/programme cycle – from attracting the opportunity round to operation and maintenance.

8 Client involvement

Whether the client wants to be 'hands-on' or not

Size of the client organisation

The client has a decision to take at the outset of a project – whether to be closely involved in the decision-making as the project progresses; or whether to set out clearly what he wants, appoint organisations that he has confidence can deliver it, and come back when it is ready for acceptance.

The factors affecting this choice may include:

It may be appropriate to have a lean client organisation in the following circumstances:

- The client is running a mainstream business or programme of which the construction project is of secondary importance; it is the means to an end, not the end in itself. In such cases the client may be too busy to be hands-on.
- The client may not have skills that relate to construction, so needs others with the right skills to drive the project.
- Construction is not core business to the client so it does not make sense to have staff employed with construction skills.
- The client has done similar projects before and has established requirements well defined, and therefore does not need to invest a lot of energy in establishing and driving the key decisions.
- The client proposes to appoint a team who have proven capabilities and an excellent reputation.
- The client decides that he wants to collocate the project team into a project office to work alongside him so they can be his 'arms and legs'.

Other clients like to set up a larger client organisation to permeate down into most levels of the project:

- The client's core business is to conceive construction projects and manage them from inception to completion.
- The project is complex and risky, with the client having unique expertise in how to understand and manage it, so needs to be hands-on.
- The client wants to 'divide and rule' because he does not trust others to

manage the bigger picture, so needs to be in control of individual packages and interfaces directly.

- The client is not sure what he wants so needs to be involved as the project progresses to be able to take hands-on decisions as circumstances unfold and the necessary information becomes available.
- There are many external factors and stakeholders that the client has to handle directly, to enable others to get on with the actual task of progressing the project without undue deflection.
- The client has not delegated responsibility of administering secondary trade contracts and supply contracts so needs to have the resources to deal with these as well as the hands on management of the interfaces and coordination between them.

There is not a right answer, so the client needs to determine the approach that is right for him and his project. But what is always important is to be clear on who is the ultimate leader. Multi-headed leadership is a route for confusion or the opportunity to play one off against another. Projects need a patron figure and project teams like to see a senior figure as the 'voice' of the project and the ultimate authority behind it.

Having determined the preferred approach, it is important to stick with it and by doing so manage the project on a consistent basis.

Don't buy a dog and bark yourself.

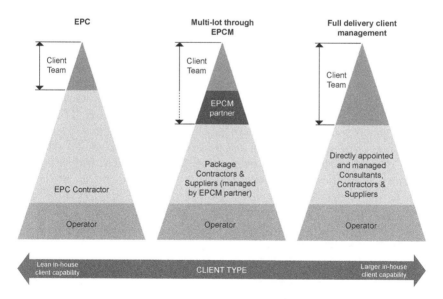

Figure 8.1 Different options for the client's structure and in-house capabilities.

Clients who build up large in-house teams and at the same time appoint external specialists to run the project create confusion and man-marking, with everyone having a vested interest to justify their particular role, even if this is just in process and box-ticking administration to check that someone else is doing their job. It may be comfortable for those involved, with everyone feeling they are managing something, but this leads to sluggish project delivery and is expensive.

Supplementing a lean client – development partner

Where the client elects to run with a lean in-house team this implies that they will be passing the project out to an EPC contractor (EPC = Engineer, Procure, Construct) or some equivalent 'turnkey' delivery arrangement. But the client has to still know what he wants; he has to have a basis of instructing this single delivery entity. There may be a case for the client appointing up-front a development partner to work closely in the client-side team, to help establish the project objectives and concepts plus quality standards. This is likely to be an independent design-based organisation with experience in creating and conceiving projects, to help with the project establishment and its definition and scoping, plus consenting and third-party approvals. In some industry sectors this is known as the FEED stage (front-end engineering and design). It establishes the concept design or 'reference design'.

This development partner role should not be confused with the project manager role. Often clients believe they need a project manager but what they actually need is the creative input to conceive the project and its definition. There is no point having someone managing something if you do not yet know what it is you want managed.

Supplementing a lean client – delivery partner

On complex projects involving multiple packages (contractors and suppliers) it may be that the client will need to be supplemented with a delivery partner (or project manager). Once the project is clearly defined, a delivery partner can help the client with the management of the project execution. This is likely to be a management-based organisation with EPCM capabilities (engineering, procurement and construction management) able to manage and administer supply chain packages and contracts on behalf of the client with appropriate delegated authority from the client.

To be most effective, the delivery partner needs to be on board after the project is defined but before the delivery or procurement strategy is established. A good delivery partner will be able to advise and take responsibility for the appropriate procurement strategy.

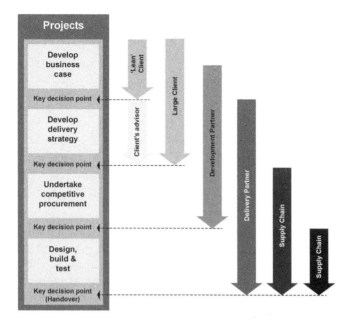

Figure 8.2   The governance of who to involve when – the 'Development Partner' and 'Delivery Partner' roles. A Development Partner can help translate the Business Case into a scheme and a way forward. A Delivery Partner can then help take the project through implementation.

These decisions on the size of the client's team and the need or otherwise for development or delivery partners will influence subsequent procurement.

9 The right procurement strategy
How best to buy the project

Introduction

The client may know what he wants as an end result, but not know how to buy it – how best to commission contracts to achieve this out-turn. His experience may or may not be in fields related to the construction industry. He may need the construction project only as a means to an end, not as an end in itself. He may therefore be inexperienced in the construction process and in construction-related contracts. The activity of appointing and signing up those who will take on the tasks of implementing the project is often described as the procurement process.

The challenge is, therefore, how to go about the project in a way that is most likely to satisfactorily deliver this means to an end. The client is interested in achieving the required out-turn performance and getting to it with the minimum of risk. The essence of a good procurement strategy is to set out and define the right process to achieve this objective. Included in this is determining the right sequence for appointing project participants.

Getting the procurement strategy correctly aligned for the project objectives is absolutely essential. Having the wrong strategy is probably the single most likely cause of projects being unsuccessful.

For this reason it is a topic that attracts considerable attention and debate from the most major of client organisations involved in construction projects. Clients with less experience should seek independent advice at an early stage of their project. The selected approach can influence the confidence in the Business Case and the ability for the project to proceed. Therefore the first steps on a project are crucial. There needs to be a willingness to thoroughly 'plan the work and then work the plan'.

Procurement strategy

The procurement strategy is the key document to define how the project is to be conducted. Who is engaged to undertake what roles and under what contractual arrangements; whether advisors, design consultants, suppliers or contractors. The procurement strategy will establish whether the project is to be set up as some form of partnership, whether some of those involved are to take a financial interest in the project, which inputs the client provides from in-house or as free-issue components or facilities, and the processes for participant selection and engagement.

The procurement strategy document should set out:

- The basis for seeking tenders, viz. full design, design and build, PFI/PPP, etc.
- The work packaging (number of contracts and work breakdown between contracts)
- The publicity of the project, sufficient to attract the right level of interest from the market
- The process for bidder prequalification and shortlisting
- The tender and evaluation process, including timeframe
- The criteria to be used for scoring and comparing tender returns
- The attitude to risk allocation and the evaluation of contingency allowances for contract packages in order to advise on the likely out-turn price, prior to making recommendation for award
- The process to obtain client sign-off and approvals to award contracts
- The structure of the tender and contract documentation
- The form of contract recommended to be used for each work package
- The allocation of design responsibilities recommended for each work package
- The timing of tendering and awarding of contract packages
- The roles and responsibilities for health and safety

Pre-engagement with industry and the supply chain

Good clients communicate with the construction industry and the supply chain well before the procurement process. By providing an early insight into their intentions and building a relationship with the market prior to procurement, they benefit from industry feedback in the development of a project, can gauge market enthusiasm for their procurement strategy, and ensure suppliers are geared up to meet project demands.

Market engagement can take many forms, including regular supplier conferences and newsletters, bidder days and, for public-sector projects, use of prior information notices in the *Official Journal of the European Union*, or equivalent.

Clients may seek to procure on the basis of securing the best whole-life value for their investment, balancing the overall quality of proposals against the cost. This may mean they choose a bid from a supplier that is not the cheapest in terms of initial cost. They also use the procurement process to thoroughly investigate the capabilities and resources on offer from potential suppliers, ensuring each member of the supply chain is right for the job.

> *An approach adopted by a major construction programme in Europe was to announce beforehand their intention to award to the tenderer closest to the median price of the submitted tenders – the thinking being that typically the lowest tenderers*

will have overlooked something or be seeking to buy the job and then be aggressive post-award, with claims to get back to a viable financial position, and that the highest tenderers are not really interested in the work, probably because they are too busy elsewhere.

Procurement as a client department

Some large clients have moved towards having a central procurement department in order to bring together this expertise. They are the team that are charged with running the process of shortlisting, selecting and making recommendations for appointment of consultants and contractors. They evaluate and score the returned prequalification and tender submissions.

This might seem logical and efficient, but it needs to be watched that the responsibility for final selection of firms to be appointed does not get solely vested with procurement. The right firm to appoint will be the one that fully understands the subtleties of the project, its objectives, technical requirements and constraints. Therefore the selection process must involve those within the client organisation that were behind these criteria and fully understand the challenges associated with the works information and site information.

However, for some reason there is an increasing tendency for 'procurement' to have become a self-fulfilling industry, as almost an end in itself. Vast amounts of effort are being put into all sorts of extravagant and complex procurement processes which appear very worthy and well intentioned, but somehow may have lost the plot. It can result in the client conveying a very commoditised approach to what should be key partnerships.

Selecting firms to appoint purely by a procurement points-scoring evaluation and largely on the basis of lowest price, without close involvement of the technical end-user side of the client, can result in what might look to be a perfectly logical supply chain appointment, but one that is actually sub-optimal. This can lead to misunderstandings and consequent misfit of expectations post award.

The best clients place the ultimate procurement decision of key supply-chain engagement at the client sponsor level; with close coordination of the advice and expertise provided by both the procurement and technical/end-user sides of the client's organisation. In this way ownership of the selection decisions is clearly vested with those in the client team who are actually directly responsible for leading the project and for the success of its supply chain relationships.

> *It does not work for the client's project leadership to be able to contend that the project did not go well because their procurement department selected the wrong firms or negotiated such tough conditions that it was a battle to get the right performance from start to finish.*

Procurement can be falsely perceived as the route to value. Many major public sector and regulated sector industries are now believing that procurement is the route to ensuring improved efficiency and value. This has been encouraged by the UK Treasury in response to justified concerns about a tendency for significant cost overruns in major capital programmes. However, this is driving a focus on procurement rather than effective out-turn.

The process is all justified by being apparently highly worthy and scrupulously fair and auditable. But it might be argued that there is a tendency to 'dumb-down' and appoint firms of the minimum standard, rather than promote and encourage the best. So the firms most likely to be able to deliver excellence and overcome challenges in execution are less likely to be awarded. If it scores maximum points with the procurement bureaucrats, why not fill the team with low-cost offshore production, even if it has no feel or understanding of the realities of delivery in the UK?

This is a flawed logic. There is almost an implication now that procurement is the end goal. The reality is that it is only a point in the process and it is the out-turn price that matters, not the price at which the work is awarded. Anyone can award a contract at an attractively low price, so procurement processes aimed at reducing the award to the lowest common denominator are likely to be 'false dawns'. Administrative bureaucrats and managers have been allowed to set processes that disenfranchise the best practitioners.

Procurement is only an intermediate point in the overall project; it is not an end in itself. It should therefore be a process involving minimum expense, time and effort in order to enable those who are best able to actually do the work to win the contracts to do it.

How much effort to spend on procurement

There seems to be a view that good procurement is about letting everyone have a go and then having a process to select out from the wide-ranging applicants and expressions of interest those who are serious contenders. This is all done through a fully auditable process to protect the client from challenge.

However, the following diagram illustrates the potential inefficiency and cost of this approach, both to the client organisation and to the industry. It is, in effect, 'job creation' for the client's procurement team as it gives them plenty to do and it is all justified by being highly worthy and scrupulously fair and auditable. But it might be argued that they are the only real winners in an unnecessarily complex process.

The diagram illustrates a not untypical procurement process (progressing from the outside of the diagram towards the central target) from first inviting expressions of interest, to then selecting a first-stage tender list of 12, which is reduced to 7 at the second stage, in order to finally select 4 firms onto a framework. But then, with the need to competitively price individual work packages through one of five different pricing routes, including demonstration of competitive pricing from their supply chains, before reaching the target of actually awarding work to undertake. All very logical until one considers the ratio of effort spent in negotiating with and working out who will not be appointed (area in light grey)

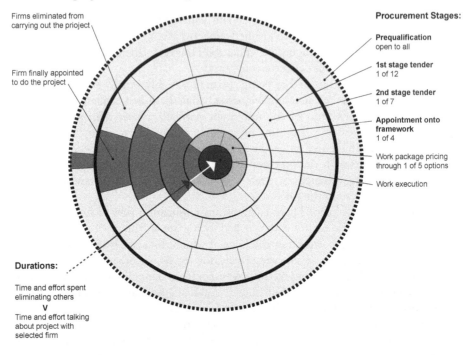

Firms eliminated from
carrying out the prioject

Firm finally appointed
to do the project

Procurement Stages:

Prequalification
open to all

1st stage tender
1 of 12

2nd stage tender
1 of 7

**Appointment onto
framework**
1 of 4

Work package pricing
through 1 of 5 options

Work execution

Durations:

Time and effort spent
eliminating others
v
Time and effort talking
about project with
selected firm

Figure 9.1   Illustrative procurement model – an example moving through various
stages towards the centre to reflect on the ratio of effort in a competitive
selection process.

relative to that spent engaging with the firm that will eventually do the work
(area in dark grey). Most of the effort is ultimately wasted effort that takes time,
costs the client a lot of money, and involves ultimately unsuccessful cost, time
and effort to the whole industry and their supply chains.

Most construction firms need to have a success rate in bidding of say at least 1
in 4. They cannot afford to carry the costs of more drawn-out processes and have
to find a way to recover these costs by absorbing them into those projects they
do win.

Procuring the right firms

Whether appointing consultants, contractors, or combinations of these through
D&B or other forms of contract, consideration should be given to the means to
attract the best firms to bid. The procurement process may be the first mechanism
by which such firms have the opportunity to see and assess the client's approach.
This will provide them the chance to determine whether the client is likely to treat
the firms he employs professionally and fairly. If they compete for work are they
doing so on a level playing field, and will the client team's behaviour through the
procurement process instil confidence that the project has a good chance of success?

Avoiding the 'procurement trap'

A good procurement strategy defines how organisations can prequalify for defined roles on the project and then the process by which any competition is evaluated to enable selection of the preferred organisations.

A key consideration needs to be the ability of the market to deliver, both in terms of capacity and capability. If the prequalification process is set up to require combinations of skills or a level of capacity that does not exist in the marketplace then the procurement strategy is flawed.

Another trap can result if the prequalification process requires firms to be able to demonstrate complex in-house management practices and procedures if these form an obstacle to the very firms with the particular expertise that the project needs from being able to prequalify. Given this procurement trap, such specialists are forced at best to be involved as subcontractors to other larger firms. This has the net effect of making the skills that can add real value to the project being at least one step removed from the client, with an intermediary who has less expertise but imposes an additional cost to the client for apparently satisfying the prequalification criteria.

So, a good test for the procurement strategy is to apply the question – '*Is the proposed approach enabling those who are best able to do the work actually to win the contracts to do it?*'

This can particularly apply when considering SMEs (small to medium-sized enterprises). They may have the very product that the project can most benefit from, so everyone is losing if the procurement process is putting hurdles in the way of them being able to compete.

> *So what is the right way forward? Keep it simple. Work out what you can afford to match your business case. Take off a contingency (do not disclose, as it will get spent if known about). Then award to the very best quality team you can afford. Incentivise them to come within their target on a 'win–win' basis, with motives that align closely with the project ambitions. Make sure you have senior level commitment. Make sure everyone can see how you want to have visibility and the ability to control issues relating to spend and programme. Then spend as much time as possible building a combined sense of team with a common purpose.*

Market capacity

Major construction projects can have a significant influence on the local construction industry. They can be large enough to distort the market and can attract international league contractors that were previously not an active influence. This may generate opportunities for the project to use combinations of skills and capabilities that were not available in the local market, but it may introduce complications of new players who are not familiar with local industry supply chains.

Most large projects need to split down into a series of packages. A key aspect of the procurement strategy is the arrangement of these packages, the scope assigned to each and, importantly, the interfaces between them. A good procurement strategy will set these packages so that they will attract strong and competitive interest in the prevailing market conditions, and so that the interfaces between packages can be successfully managed. Awareness and preconditioning of the market so they can anticipate and plan for the inputs that will be expected is a further consideration when mapping out the project procurement plan.

Feedback from procurement

Good clients provide feedback to their supply chains on their performance through prequalification processes and using key performance indicators. They also encourage suppliers to give feedback about any issues that may have arisen.

Good clients sustain the open and clear communications they have developed with their supply chains after they have been contracted to work on a project. Confidentiality is the exception rather than the rule, and such openness ensures any concerns about delivery are highlighted early on.

Another aspect of procurement is the 'quality' versus 'price' ranking. Very commonly clients insist that they are making judgement with transparency based on both. They even often argue that they have 'weighted' the procurement department's scoring so that Quality might be 70 per cent and Price 30 per cent of the total marks. However, slightly more detailed scrutiny can show this is a convenient but often flawed premise. Usually the Quality score is a bundle of separate issues such as technical, track record, team structure, diversity, safety record, and so on. Each aspect is given a sub-total and often the resulting spread of scores between the competing firms is quite small. The Price can then be scored as a single issue in an almost arbitrary way to create a spread that completely outweighs the spread in the Quality score – for instance, the lowest price getting full marks and the highest zero, with the rest spread in between. In effect this results in combined assessment and selection being almost solely price-driven. It warps the scoring so that Price dominates over Quality. The only way to fairly reflect a stated intention of a 70:30 weighting would be to moderate both sets of scores so that they have the same proportional spread. However, this is often not done.

Gateway approvals or continuous process

Some projects can be treated as a single continuous process through to completion once project authorisation is granted. Others are required to go through a series of progressive approvals or 'gateways' in order to receive authorisation to progress onto the next stage. If significant funding is required from external financial institutions then a key milestone for project authorisation will be Financial Close – the point at which the project reaches sufficient definition that it is confirmed that the required financing is in place.

The commitments that can be given are very different for the two scenarios and the Procurement Strategy must understand and recognise the conditions appertaining to the project. When going to the market there is a need for clarity on the level of authorisation that has been achieved and the current objectives and committed work horizons and timescales.

It has already been stated that successful projects require the objectives to be clear and visible. By definition if a project is required to go through a number of gateways then the ability of the project as currently established to achieve the final end objectives is not yet accepted and there is not yet total client commitment to the project. The consequence of this is that if the client cannot be committed to a defined end objective, then he cannot provide a clear picture and commitment to others. The result is a reluctance for prospective tenderers to bid or, if they do, the bids are likely to be uncompetitive.

Although introducing multiple gateways into a project may be seen as improving project governance, it may actually be counterproductive since it becomes very difficult to build the required project commitment and momentum from those who are needed to make it happen. Gateway reviews need to be watched to avoid the next 'gateway' being an end in itself, with the consequence of short-term horizons consisting of process and box-ticking that seem to imply good governance. The danger is that this can result in the whole project not being seen from beginning to end. Successful projects tend to be 'pulled' to a clear objective, not 'pushed' from one milestone to the next with uncertainty and scope for change at each. Also gateways should not be points at which the primary project objectives can keep being changed or backtracked on.

If major gateway approvals are essential, then it may be better to consider each stage from one gateway to the next as being a project in its own right. Then the objective can be set unambiguously as the task of achieving the next gateway approval.

Interface considerations

Experience is needed of awarding and successfully managing contracts with various combinations of work scope, selected to suit the particular circumstances and availability of appropriate skills and resources. In some cases it is beneficial to award complete all-inclusive packages. In others, where the work has to be installed in a complex sequence of stages with external interdependencies, it may be more flexible to split the scope. Some elements of a project may require consistency on a project-wide basis, whereas others may be more suited to a geographic subdivision of the project.

The procurement strategy should seek to optimise interfaces in order to:

- Minimise the number of interfaces between contract packages (whether design-design, design-construction, construction-construction, construction-commissioning).
- Place and define required interfaces between contract packages so that they will be relatively easy to control and manage.
- Introduce specific, managed interfaces, where this will permit an increased

confidence in the successful delivery of the overall project, or an improved control of the overall out-turn cost.

- Place interfaces between work packages to the benefit of the project where this would permit early work packages to be tendered and awarded to suit the overall project programme, whilst allowing other more complex packages to continue to be developed in parallel, so as to better define their scope and/or mitigate potential risks.
- Enable different contract conditions to be used to best suit the different elements of work.
- Enable the selection of 'construction only' for some contracts and 'design and build' for others.
- Allow contractors to be selected to best match the skills and capabilities required for different workscopes and thus achieve improved confidence in their performance and more competitive pricing.
- Enable risks to be allocated differently between client and contractor for different work packages in order to place risks with those organisations most able to manage them and thus control the out-turn price and the programme.
- Split or subdivide the work so that more contractors have the ability to competitively price for it, to spread the risk of delivery and/or to enable smaller contractors with lower costing structures to undertake elements of the work cost-effectively.
- Understand and define who owns programme float at interfaces.
- Arrange packages of work so that they can be commissioned and handed over early, either to allow follow-on contracts to be able to proceed cleanly, or to enable the commencement of operation.

When setting up interfaces it is important to be clear who it is that will be responsible for decisions. If the two sides of the interface have slightly different ideas that require coordinating, who is it that makes the call? Is it one of the two who is contractually responsible for this, and if so how are the consequences on the other accepted? Or is it a client or client's project manager that carries this responsibility to make a timely decision and instruct both sides of the interface? A thing to watch in this is that many project managers are not set up to take technical responsibility so are reluctant to make technical interface judgements.

A further important factor is the programme/time aspect of interfaces. It is necessary to anticipate this and be clear on the definition up front of who owns programme float, and particularly 'terminal float'. If one contractor completes early at an interface, the client needs to avoid having to 'buy out' the terminal float in order to give the benefit to the succeeding package.

BIM and shared electronic data platforms

There is one aspect of interfaces in construction that is rapidly evolving and changing. The increasing move to three-dimensional electronic models and databases means that there can be a complex single model to represent the whole project. All involved can access this model and contribute to it, with the great advantage of there being *a single version of the truth*. Also the advantage of being able to coordinate and integrate the various aspects of the design. These models are then moving to four and even five dimensions, with the inclusion of cost and programme information.

However, the challenge here is who 'owns' the model and the information in it, and who takes responsibility if all parties involved in the project are able to contribute to it and make additions or changes to it?

Many of those involved in projects used to preciously guard their design inputs as their intellectual property (IP) and control of information. This may have led to more fragmented information but did have the advantage of contractual clarity over responsibilities for accuracy and completeness.

There is an inevitable progress towards 'digital workspace', virtual information platforms and the seamless transition from CAD (computer aided design) to CAM (computer aided manufacturer) to RAMS (reliability, availability, maintainability and safety) project records and operational control. It will require clients to consider equitable contractual arrangements that support a partnering approach to project information. This may have quite a far-reaching impact on professional liabilities.

Planning consents and other third-party statutory approvals

A major consideration in developing the procurement strategy and in defining the work packages will be the ability to achieve planning consent and other statutory approvals. Proceeding without approvals in place is a major risk and one that contractors cannot reasonably be expected to manage. They may be expected to manage and close out the details of 'reserved matters' and other conditions associated with the consents but not the overall risk of gaining approvals. If work packages are contracted before there is adequate close-out of consents and approvals then any material changes that are required to satisfy the authorities provide the contractor with a reason for variation and delay. This can be avoided by packaging the work so that workscope for which adequate approvals have been obtained can be progressed and contracts awarded, whilst the consents for other later packages are still being negotiated.

This approach means that works contracts are not awarded until the planning consent and other statutory approvals are adequately closed-out. This avoids there being an associated risk and scope for uncertainty against the awarded work. A good procurement strategy will reflect this.

The appointment of the project manager

The client needs to determine whether he has the resource and skills internally to manage the project. The client must have a senior individual identified as the project client (or project sponsor), who is empowered to take the strategic decisions. However, the project client will require support from a management team able to handle the practical administration of the project and to take responsibility for delivery of the project, operating within the client's strategic framework. Increasingly client organisations do not have such capabilities in-house so appoint an external project manager.

Logically this appointment is made early, once the business case and governance proposals for the project have been formulated. It is clearly very important to have a project manager that has a close empathy with the client and the client's values. Equally, it is important to have a project manager that understands the distinction between the development and delivery phases of the project.

The appointment of consultants

The consultants are the team of specialists who are able to prepare the design solutions for the project. If there is a need to develop the creative ideas that will define the project then consultants will be needed early and in a close, direct appointment with the client.

The need to gain approvals, planning and environmental consents and the need for design development to achieve confidence in project pricing, relative to the need to achieve single point responsibility for project delivery, can be factors in whether the client elects to engage consultants directly or to have them appointed within the larger project delivery team.

The client may elect to appoint a single combined team to fulfil both design and delivery functions. However, if the consultants are part of a larger delivery team then consideration needs to be given of how the client will be able to assess that he is getting what he asked for. In some cases clients elect to have the single point responsibility benefit with the consultant being appointed by the delivery team but then appoint directly a separate consultant, independent of the delivery team, to fulfil the role of client representative/technical representative. The downside of this is that it does result in some doubling up of consultant costs that are ultimately borne by the project.

The appointment of the contracting organisation

The procurement strategy needs to identify how and when the contracting organisation is brought onto the project. There are numerous forms of procurement that have developed over time for use by construction clients. As with the selection of the form of contract, there are advantages and disadvantages in the different options and the client needs to understand these to be able to make an informed decision.

Factors to consider or amplify in considering the right option include:

- Timing of obtaining a contractual commitment and binding price
- Need or otherwise to overlap design with construction
- Need for single point of responsibility

- No ambiguity over safety and the degree to which this is delegated down the supply chain
- Need for contractor input into complex construction sequencing
- The spend profile between the design phase and the construction phase
- The complexity of design approvals and sign-off before committing to construction
- How the project is funded
- The relative importance of cost, programme and quality
- Who is in control of quality and design excellence

The best clients evaluate the requirements of their own project and the results of any market research they have undertaken, in order to establish the most appropriate form of procurement to use. They then communicate their decisions on procurement strategy to the supply chain at the earliest opportunity.

Choice for form of tender / form of contract

There are a number of different approaches to contracts (see Figure 9.2):

- Fixed-price lump sum
- Fixed price with re-measure on completion to an agreed schedule of prices to pick up any variations
- Two-stage contracts, or framework agreements
- Target price with payments made against actual costs
- Incentivised target price with actual costs and 'gain/pain' arrangement

The procurement strategy will set out which type of contract is recommended for each of the packages. The choice of contract will be determined by the nature of the works. However, best practice generally means using a contract that encourages collaborative working. The UK Office of Government Commerce recommends NEC3 for this purpose. (The NEC suite is the New Engineering Contract form promoted by the Institution of Civil Engineers for use on major construction contracts.)

Fixed-price lump-sum contracts work well and produce the lowest costs for relatively straightforward work packages in which the scope and the risks are well defined up front. They are not cost-effective where the risks are significant, with dependencies upon factors outside of the control of the contractor; either contractors will not be willing to accept these risks, or they will put a high cost allowance into their price, which the client will have to bear whether or not the risk materialises.

Fixed-price with re-measure contracts move some of the risk back to the client, by accepting that variations to the work scope quantities will be reviewed and reimbursed.

Two-stage or framework call-off type contracts are appropriate when the work scope involves technology development or scope development that is not able

	Description	Pros	Cons
Traditional	Contractor appointed to deliver a defined set of construction works	Client retains design control Relatively simple and quick Can achieve good price certainty	Potential for conflict and delay due to interface between designer and contractor, particularly if design not complete at contract award
Two-stage	Contractor chosen on basis of limited information before negotiating final deal	Can bring contractor on board earlier, speeding up delivery	Can be difficult to ensure best supplier is chosen on basis of first stage tender and second stage pricing may be difficult to control
Early contractor involvement	Contractor appointed early, often in parallel with the design team	Contractor's early input provides constructability input and brings strength in negotiations with supply chain	Can be difficult to get the contractor to price competitively or accept risks and they may use their early involvement to avoid exciting options if thought challenging
Construction management	Appoint construction manager to oversee works, delivered by trade contractors in direct relationship with client	Can secure rapid delivery Allows development of relationship between client and key trades Increased design flexibility	Client retains significant risk of cost increases and interface risks between trade contracts
Framework	Appoint supplier on basis of a pricing structure, which is then applied to various projects over the life of the contract	Reduces the requirement to procure on a job-by-job basis where client has programme of works Allows client and supplier to benefit from continuous improvement over a series of similar projects	Difficult to ensure framework is continuing to provide value-for-money over life of contract

	Description	Pros	Cons
Design and build	Client appoints supplier to deliver both design and construction of project	Client deals with single delivery team so single point responsibility	Expensive to bid so can restrict contractors willing to submit tenders

Some loss of client control over design compared with traditional as designer is employed by the contractor |
| **Design Build Finance Operate**

(e.g. Private-finance Initiatives, Public-private Partnerships) | Public sector client appoints consortium to fund, design, construct and maintain building or structure for long term concession | Long term contract incentivises supplier to deliver whole-life value rather than cheapest build cost

Risk transfer from client to supplier

Deferral of costs to primary client body | Not suitable for all but the largest projects

Client loses significant control of project

High cost of setting up PFI concession or financing arrangements so overall cost is higher |

Figure 9.2   Forms of contractor procurement.

to be adequately defined at the outset, or where there is a need to negotiate a general commitment to a specialist contractor or supplier so that they can be used for a series of packages progressively on a project-wide basis to ensure a system-wide consistency.

Two-stage contracts need careful consideration as in the second stage they become in effect single-source negotiations with a high level of pre-commitment. Attention therefore needs to be given during the first stage at the point where there is still competition to achieve adequate tie-down of the basis for the second-stage negotiations.

Key factors to consider at the first stage are:

• Overall project scope needs to be defined so that the management scope and range of trades can be understood.
• The allocation of design responsibilities between designers appointed to the client and design to be undertaken under the contractor's remit needs to be defined, as this can have a big impact on the contractor's overheads.

- Must fix rates for pricing against an appropriate order of magnitude of quantities, so that there is an agreed basis for the second-stage finalisation of prices once the quantities are more clearly established.
- Principal contractor responsibilities for site security and safety.
- Overheads, profit and preliminaries need to be established.
- Project programme must be established as the context for the overhead and preliminaries because these are time-dependent costs, so there can be significant growth at the second stage unless they have been tied to a programme at the first stage.

Early supply chain involvement may be considered, often many months or years before work starts on site. Such 'early contractor involvement' can be beneficial in enabling clients to take advantage of a supplier's expert knowledge during statutory procedures and planning, or of buildability and cost estimating. These may allow a better understanding of the project before work starts so could contribute to accelerated delivery.

Target-price contracts involving the tracking of 'actual costs' generally are good in encouraging a commitment to achieving the project programme. They also encourage a flexible approach so can be good in enabling the best management of complex risks. However, they are administratively intensive for both contractor and client, so carry a high administrative cost. This form of contract tends to be most appropriate for large complex packages. Where there is an incentivised 'gain/pain' arrangement, careful consideration needs to be given to the independent control of quality.

Incentivised target price with actual costs and 'gain/pain' arrangement

Target price contracts usually involve a relatively sophisticated understanding of allowable actual costs which can be tracked relative to the target. How this is applied to the supply chain needs to be understood and agreed up front as much of the construction industry supply chain may not have the ability to truly track actual costs.

Incentives may seem like a good idea, but should not be entered into without a real understanding of the behaviour they will drive, so need to be carefully considered. For instance, do they relate to both cost and programme? If so, who 'owns' programme float, particularly terminal float, and which party benefits or loses if programme float has to be used or reallocated? A similar logic needs to apply to cost contingency. An obvious point, but one that is often overlooked, is that if incentives actually occur at more than one level (linked to multiple KPIs) then there will be a vested interest for individual parties to work out which of the incentive mechanisms best rewards them and skew their focus accordingly. This may actually influence collaborative behaviour to the overall detriment of the project.

There is a tendency for 'incentivised' contracts to be associated with the upside of bonus gains. But most incentivised contracts include a 'gain/pain'

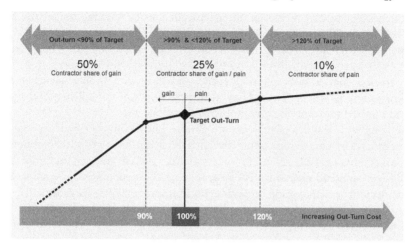

Figure 9.3   Incentivised Contract – illustrative 'gain/pain' mechanism

mechanism so there is also a potential downside. Setting the upside gain and the downside pain does not need to be a straight line. It can influence behaviours. Considering this further, if the client is clear enough about its business case and can see it can generate the required business performance and profit if the project is delivered for the target price, then any improvement on this is a straight bonus to the client. If it is already going to make good commercial returns it can afford to be quite generous to those helping make this possible. The more they benefit, the more the client further benefits, so this is well worth encouraging. It needs to be given as much focus as possible.

By comparison, what happens if things do not go so well and go over budget? This is clearly less desirable to the client as its commercial business case margins will start to reduce. However, to contain the overrun the client needs all parties to still be putting most of their efforts into getting the project finished. The client does not want the contractor's primary efforts being switched into building a big case for protective claims. Or even worse for the contractor to look for a means to walk away and leave the project unfinished. This logic tends to advocate different share margins for 'gain' and 'pain' in order to drive the desired behaviours for the best advantage of the project.

Integrated supply chain

Having appointed their supply chain, experienced clients draw them into the project team and aim to develop relationships with all parts of the chain. By integrating the whole supply chain, clients can make good use of the skills of individual suppliers for the benefit of the project. It also ensures that any problems arising in more remote parts of the chain are known and resolved more quickly.

Such collaborative working, often with all parties working out of the same offices, will result in people working together to deliver the project rather than merely looking after the interests of their own organisations. Often clients achieve success by instilling a 'no-blame' culture within the integrated team, recognising that it is better to identify and deal with problems rather than unknowingly let them fester. They also provide a well-established and understood method for resolving any problems that do arise.

Clients with a series of projects are able to harness further benefits that can be achieved by continuous improvement. They agree targets with suppliers for improvements over the series of projects in areas such as the cost and time taken to deliver standard elements. They can expect their suppliers to demonstrate a commitment to innovation to secure these improvements. In return, the best clients reward well-performing suppliers with continuity of workload.

Importantly, such high-performing clients also agree and adhere to a transparent structure for payment for all works carried out. They see benefit in setting up project bank accounts. As part of this process, they may look to incentivise suppliers to achieve the best possible results. Equally, they may include a way of ensuring that any losses occurring on the contract are shared across the project team. Any such 'pain/gain share' arrangements are prepared on a fair and open basis.

Following completion of works, clients should aim to make payment promptly and ensure that this payment is passed down to all members of the supply chain. Experienced clients understand that they attract the best suppliers if they avoid stretching payment terms, or withholding payment through the use of retentions if the work has been carried out to the agreed specification. A procurement strategy that is unfair to those in the supply chain may work once, but clients who repeatedly achieve success have recognised that procurement policies that treat others firmly but fairly yield benefits overall to the client.

Procurement on 'quality' or 'price'

The transparency and interpretation of the client's objectives on 'quality' vs 'price' is worth further consideration. As explained earlier there is a tendency for the client's procurement process to set out claims of a strong emphasis on quality (viz. 70% on Quality and 30% on Price), but the reality can be somewhat different. What needs to be understood is that the spread of scoring under each of these can be the crucial determining factor. If the spread ratio is not comparable in the total scoring, then the resulting rankings mean that the emphasis is anything but transparent. The driving factor can very easily be skewed to be 'price', suggesting the stated 'quality' focus was not really the client's objective.

Top-quality firms want to compete on a level playing field and to know that they are pitching against others who are in relatively the same bracket within the industry. They do not want to be bidding against the low-cost commodity end of the market. If the client's driving motive really is top quality then a more transparent procurement approach is with 'two envelopes'. First pick the winner on quality and only then open their price envelope. If this is acceptable within the context of the client's business case then there is a basis for appointment on the grounds of best quality. If the price is not acceptable then have a negotiation to see if their price can be reasonably adjusted. Only if this does not work, go to the price envelope of the second on quality.

Matching technical specifications to the form of contract

Technical specifications need to be compatible with the selected form of contract. Often the general sections of specifications define responsibilities for submissions for review and approval and set out the process for the acceptance of quality and approval timeframes. Contracts such as FIDIC (International Federation of Consulting Engineers) or NEC (Institution of Civil Engineers) have different forms (or 'books') in their suite of procurement options, ranging from 'employer design' through to one-stop EPC. The specifications must be aligned to suit.

10 Forming the project team
The relationship wanted with the project team

Introduction

The client needs to consider whether he wants to involve organisations he has used before or whether to go out for open competition, with the possibility of this resulting in having to form new relationships.

For some clients, particularly those in the public sector, this is not an open choice since they are obliged to comply with procurement regulations demanding competition and equality and diversity. Such requirements may give the client little choice or may even dictate his whole procurement and contracting strategy.

Issues to consider

The client may want to achieve a collaborative partnership with his project team.

Or the client may be clear that he wants a 'contract giver' and 'contract receiver' relationship in which there is no doubt over who is working for whom.

Good fences make good neighbours.

The strongest partnerships or teams are often formed by team members having clear areas of expertise and clear roles. If these are mutually supportive there are the strong ingredients for success. Everyone trying to do everything and believing they have a say in every decision may be a route towards inefficiency and confusion.

A good client will recognise that major projects over a long time span need to permit individuals to develop and progress. He cannot enforce upon his team that all individuals are bound into fixed positions indefinitely with contract conditions that try to enforce this. Such constraints are counterproductive as good upwardly mobile people need to be able to progress. If the project is correctly set up, then these people will behave responsibly and actively develop others, to the benefit of both all individuals involved and the overall project. The project can thus achieve a level of influence and a reputation within the industry as a breeding ground for excellence.

The attitude to safety can be hugely important. Given the right commitment and conviction to actually make safety everyone's duty from the top down and bottom up, rather than somebody else's job, the project can gain a reputation of being a 'good place to be'. This matters not only to the project team but to their families and dependants.

Project people

The success of a construction project rests with the team charged with its delivery. Clients need to appreciate and understand the mutual respect needed to cement project team relationships, and the importance of engaging the most appropriate project people.

> *Generally up to eighty percent of project success is attributed to what people do and how they do it. The rest is attributed to the other matters covered by this guide. The challenge is that most of our effort goes into the other matters, with the people matters considered to be so called 'soft issues' which can be dealt with tomorrow – but tomorrow never comes.*

The attributes of those people engaged by the client organisation to lead the project team are worth detailed consideration. Smart project people drive best practice and feed back into the profession and industry. They are result-focused, strategic thinkers, rigorous analysers and planners, and pragmatists. They engage with and empower others. The competencies required for success are a combination of technical and behavioural qualities, and require a high degree of emotional intelligence.

The best project people propel the project to its end destination with passion. They generate excitement around project goals, and drive and energise other members of the project team to deliver consistently high standards and results that exceed expectations. These project people constantly challenge constraints and seize opportunities, often in what would normally be perceived as adverse circumstances.

Creating clarity – As with any team, there needs to be the right balance between 'creators' and 'deliverers', but good project teams have people in senior positions who instinctively simplify complexity and create clarity. By analysing and integrating complex, often ambiguous issues and maintaining the strategic view, they are able to make informed decisions. They are broad, innovative, long-term thinkers. By anticipating external developments and scenarios likely to affect the project, they identify innovative solutions and ways of working.

Rigorous planning – In the delivery role, project teams need people who are capable of analysis and judgement. By rigorous planning, identifying trends, links and themes in business data, and using complex links to reach conclusions and make sound decisions, they take calculated risks to add value. They are also skilled at business and programme planning, able to coordinate, organise, direct and implement a portfolio of activities.

Pragmatists – Project people spearhead a high-performance culture and attitudes which encourage collaboration and learning opportunities. By creating and leveraging internal and external partnerships, they ensure issues are pragmatically resolved before they become problems. By building strategic networks around the project, project people can manage multiple and complex relationships.

Balance – The client and project team need to achieve a balance between the attributes of the project people and the nature of the project, over the life cycle of the project. For example, strategic thinking at the feasibility stage of the project is crucial, but during commissioning and handover, pragmatism and result-focussed attitudes will be required. Equally, pragmatists can often scupper complex amorphous problem solving during the delivery of the project by reaching conclusions prematurely and not accepting the advice of the project team and its advisors.

> *You can have systems and governance processes, but great projects are made by great people.*

Assembling a project team

The most effective clients understand the culture of the project workforce, which means understanding how project people are motivated and how they gel as a team. The way the team works is crucial: a loose grouping of individuals with different objectives will not deliver effectively.

Nurturing a high-performing, empowered team, and one that changes through the life cycle of a project, is crucial. Investment in early and repeated team-building events away from the routine activities of the project is hugely beneficial.

Not all 'project people' are of a particular type, but there are some important traits that distinguish them from their 'corporate cousins', who tend to internalise and have much longer time horizons. Of course, there are people within large organisations who are clients and project managers, but they are often the minority, and often oversee either part of a very large project, or a project that has several senior responsible owners over several years. Understanding this helps clients assemble and work with project teams to deliver projects successfully.

Some characteristic traits of people who succeed in a project environment are as follows:

- They can adapt to and more readily accept change than their corporate counterparts. They have a positive, can-do attitude and good judgement of appropriate use of time and prioritisation. However, they are less keen on mundane tasks and processes.
- They are motivated by tangible delivery rather than organisational or corporate aims, and often have deep-seated views on social responsibility and the impact of the project on the community.
- They enjoy challenging environments and generally a fast pace of work. They quickly rise to a challenge and feel rewarded by execution of the resolution.

When unable to gain promotion on any one project, they will often force the pace of 'on the job' learning to grow in competence for the next project opportunity. Good progress encourages more progress with these people.

- They are natural team workers. On well-run projects, team working extends into the social as well as work environment, building bonds further.
- They are attracted to self-improvement, with a ready eye on the market. Their mobility tends to be greater than the norm, as does their flexibility in accepting short-term assignments that will give them an advantage in the long term.
- They are driven and pragmatic, but generally well rounded. They tend to have engineering, construction or programme management training, which is heavily enshrined in logic. Working with fellow professionals, often under intense pressure, reinforces the pragmatism, but the balance is often found in listening to and respecting the opinions of others.

There is never a perfect project team, and some of the above traits may equally apply to other types of organisations. The point is that recognising these whilst recruiting and building the team will lead to a better chance of high performance, particularly when there is little or no time for specific shaping and training to develop individuals.

> *What style of leadership? It was tough and single-minded, but recognising that the project would only be successful if we created a team spirit across all the sectors, even when we were arguing with contractors, banks and other stakeholders, it was the team spirit that said, 'We are going to do it, that is everybody was going to do it.'*

Co-location

Clients increasingly consider whether or not it will be to their advantage to have the team for major projects co-located, working together in a combined office. This debate is applicable for many significant projects in the UK, but is also relevant to working internationally. In both cases the driving concern is assembling a committed team that can be relied upon to deliver the project without being sidetracked onto other conflicting priorities.

International clients, such as those in the Middle East, know that the normally available local resource is insufficient to be able to deliver their large projects. They therefore have a tendency to ask for a commitment for consultants to place their team locally; in effect to co-locate the team in country.

Clients need to attract the best project talent available and retain this for the duration of the required roles in the project lifecycle, but it may be that the inflexibility of being co-located makes it difficult to attract the best people.

On occasion the reason to co-locate is to form a seamless collaborative partnership. In others the requirement is the client wanting to exert power; to assert the 'contract giver' and 'contract receiver' relationship in which there is no doubt over who is working for whom and the client wants this co-location to reinforce the sense of control.

The question therefore is whether co-location creates true benefit or whether other factors mitigate or offset the potential advantages.

Team co-location considerations

The following are some of the principal issues to be considered when making a choice on collocating the team.

Achieving a common agenda

Pros: Co-location and the formation of a single team is undoubtedly a powerful benefit in building a single objective and a common understanding of the objectives and project priorities.

It enables the client to have a hands-on presence and real-time influence. Issues can be identified early and the best combination of people asked to find a solution. Commonly staff enjoy working in a strong team environment, in which work priorities are reasonably easy to establish.

Cons: This may not work if the agenda for the project is still not determined, if the procurement strategy for key participants is not established, if the project is required to go through approvals stages without confidence of proceeding further, or if the inputs of the different participants are needed sequentially rather than simultaneously.

Client staff involvement

Pros: A big gain is to enable the best of the client's staff to have a clear influence on the project and to help establish a team ethos and instil the prevailing attitudes towards what constitutes success.

Cons: This needs an attitude of sharing in which the person most able to take a problem and solve it is willing and encouraged to do so. Co-location is unlikely to be the right answer if the client wants just to push problems 'over the fence'.

'Project team' vs 'company spirit'

Pros: Co-location is usually a good way to get people to really understand one another and to build a collective team spirit with the mission to achieve a successful project.

Cons: Dislocation of staff from the wider influences within their parent companies and the accessibility of their top leadership can result in the special values associated with a company being diluted. The very reason for having selected a firm for its unique qualities may be lost.

Single leadership

Pros: A single leadership team can be formed with the collective mission to drive the best interests of the project, regardless of the particular interests of the participating companies.

Cons: This objective needs careful consideration as there needs to be a willingness for individual participating firms to subordinate control over their own staff. Some of these firms may have been selected because of their unique off-the-wall creativity and this may be suppressed if other cultures are more dominant through a single leadership approach. However, to have multiple leadership of a collocated team is likely to result in rather visible chaos and disharmony.

Size and nature of accommodation needed

Pros: The project office can be a signpost providing a clear identity for the project, its importance and sense of purpose. Facilities can be bespoke for the needs of the project and it can be located to best suit the project and to minimise travel time and the availability of participants for meetings.

Cons: The objective when forming a collocated project team is to achieve commitment, a common bond and, above all, stability. To make this possible it is desirable to avoid the need to frequently relocate the team. This can be expensive as there needs to be a real allowance upfront for the size of accommodation required, anticipating the various stages of the project. If staff remain located within their parent companies, the ramping up of work on one project is compensated in accommodation needs by the reduction in staff on other projects, so this enables a dynamic management of office accommodation that is not possible for a single collocated project. This control and provision of office space therefore moves from being the problem of the individual participants to being that of the client, with inadequacies being a major scope for claims against the client.

Common support functions and establishment costs

Pros: There is the opportunity for the client to take more immediate control over the establishment costs, set-up costs and overheads. An effective collocated team needs consideration of HR, project finance and controls, timesheets, IT, GIS/BIM systems, induction, training, media relations, house management, PI and other insurance, health-and-safety responsibilities.

Cons: Significant up-front planning is needed by the client, or someone appointed by the client, to achieve this. These activities all become the client's responsibility to organise and provide with a collocated team, whereas they would usually be the normal business activities of the participating firms. There can be some offset of these costs if provided by the client, but since each firm will still need the core functions for its ongoing business they cannot be fully offset.

Incentivisation for success

Pros: Common objectives and goals can be set with all participants aligned to achieve them, with a common understanding of the priorities and emphasis. The strength of the overall team will tend to compensate for areas of weakness in order to hang on to the overall targets.

Cons: Co-located working will tend to expose and emphasise any fundamental differences in the way the various participants are rewarded. It will cause difficulties if the incentives for members of a co-located team are not aligned or are disparate.

Percent billability and cost to get commitment

Pros: Staff resource is known and visibly available with a single focus on the project. Staff can be directed to suit immediate project priorities without possible conflicts from other projects.

Cons: Staff need to be fully committed and thus have to be 100 per cent billable to the project if they are co-located and this can be expensive.

Staff growth and promotion

Pros: Clients can believe that they have secured the right staff and can guarantee their ongoing input by having them co-located.

Cons: The best staff are upwardly mobile. They are attracted to the most prestigious projects. However, co-location on projects with a long timeframe can lead them to believe that they will be overlooked 'back at base' in the promotion stakes. Unless this is proactively guarded against it can result in co-located project teams not getting the best staff.

Software systems – licenses and running large analytical packages

Pros: Through a co-located approach it is normal to plan and set up single project databases and common systems. This takes effort and clear decision-making but then has the big advantage of all parts of the team working using a common, consistent basis, with information that can be controlled for its status and approval.

Cons: Licensing of specialist software may normally be part of a company overhead and such licenses will be shared across many projects. Similarly, some large packages, particularly analytical or network modelling, are run within firms by using hardware across their full network. It may be difficult, or not cost-effective, to achieve this within a co-located project environment.

How much to co-locate

Pros: Co-locating is best for the major core element of the project team that has a long-duration input and which holds the vested knowledge and intellect of the project. Co-location can be good to enable a better and wider understanding of the interests of specific disciplines, so tends to lead to better integration of multi-disciplinary solutions.

Cons: Major projects tend to have peaky demands for certain activities, such as detailed production. It may be that this work can be better undertaken using more cost-effective remote production centres spread geographically within the participating firms.

Summary observations on co-location

Overall there can be benefits from co-locating key parts of the team on major, complex projects of significance over a reasonably long timeframe. However, it tends to be an expensive option with significant up-front costs, so the client must be convinced that the longer-term benefits will be realised and must plan from the outset the involvement of the participating firms so that they can see and support the overall objectives of forming a single team.

One area that can lead to misunderstanding is the distinction between 'Co-location' and 'secondment'. Co-location in this context is the decision to form a clear project identity and to bring the various participating organisations employed by the client together to form a single entity. It is about merging the best of the various participating firms' capabilities and cultures to achieve a project culture that is greater than the sum of the parts. In some co-located teams it may also involve bringing in some of the approval granting bodies, with representatives of these asked to focus solely on the one project. This is different from secondment, in which the client merely seeks to augment his own team with staff from other organisations hired in as individuals to take up posts within the client organisation.

Attraction and retention

Clients need to attract the best project talent available and retain this for the duration of the required roles in the project lifecycle. There are specific issues that clients take into account when attracting and retaining smart project people.

There is a tendency for types of projects to attract specialists or require people with particular skills, so clients need to have an awareness of the number of similar projects that will be underway in the proposed timeframe. Networking through project management organisations and professional institutions, as well as government agency working groups, can augment specific knowledge of market conditions and resource availability. Project people also use their networks well to move from one job to another, so clients need to consider seeking the views of others as to where and how to locate the best personnel prior to launching a formal recruitment campaign. The positioning of the project is important in this respect as high-prestige projects tend to attract the highest calibre of talent at the top echelons. This in turn can be good for attracting equally good, but less expensive, talent into the lower ranks, as they see great benefit in being associated with high-profile projects, the top people running them and the internationally acclaimed reputations of the firms involved.

In a competitive market or in remote locations, clients may need to consider endorsing non-standard working hours, gym or club membership, health and well-being packages, good training and development, and extended holidays as part of the overall remuneration packages, or flex to suit individual requirements to get the best mix of project people.

Project people are often so focused on the job in hand that this can lead to them neglecting their personal development. This has the upside that they become extremely good at what they do, but the downside is that some may become disillusioned and less effective than they might be if they realise they are locked into a 'career freeze' through their commitment to the project, particularly if of long duration.

Clients should avoid contract terms that permeate this and are advised to facilitate measures to

- encourage career development of project staff, such as membership of professional bodies;
- openly encourage project staff to plan their next assignments;
- particularly encourage internal promotions on merit in long-term projects;
- encourage mentoring and exposure of younger staff to making presentations;
- promote internal training sessions for the improvement of technical and soft skills.

> Q: '*Do you think there is any room in projects and programmes in the fast-moving change culture we experience to develop people along the way and develop their confidence?*'
> A: '*Oh yes, I think projects are the best way of developing people. There is a whole industry of people to develop. {...} I'm talking about people thrown in at the deep end to deal with the varying challenges of a project or a programme. {...} Being allowed to and expected to make mistakes, as long as they respond smartly, recover and make sure they're not making the same mistakes again. It's a marvellous opportunity and that is why I have some excellent people, because they have so much practice in the really good, high-profile challenges. Very worrying if they are the worrying kind, but once they have done a few, they begin to get a lot of confidence, it is a marvellous laboratory for developing people.*'

Clients should consider encouraging mentoring and training to leave a 'soft legacy' for the project. By promoting and capturing best practice during the project, it can be quickly distilled and shared with others to ensure the further adoption and continuation of the principles developed. Mentoring and coaching across the project team can provide an opportunity to share concerns and issues outside the normal workplace environment. This often enables issues to be resolved rather than being allowed to fester and have a negative impact.

A further 'soft legacy' of real value can be achieved by schools-liaison programmes and work placements for the local community. These do not generally cost a lot in terms of project finance, but do promote considerable goodwill towards the project and provide genuine career opportunities.

11 Managing uncertainty
How best to align risk, cost and contingency

Introduction

The objective is to maximise the likelihood of delivering the completed project *'on time, within budget, safely and to the required quality'*. However, no large construction project is without risk so the consideration and assignment of possible risks between the parties involved in the project can be a significant factor. It can influence behaviour and the management of costs and time.

There has been a tendency for clients to believe they can pass all risks down to others by writing legally watertight contract terms and conditions. But if by doing so they are passing risks to organisations that are not in reality able to manage the risks then this may generate future work for arbitrators and lawyers, but is unlikely to be in the project's best interests.

Allocating responsibility for risk

A primary objective behind a good procurement strategy will be to anticipate risks and manage them by assigning them to the organisations best placed to control and mitigate the risks.

This strategy to control and minimise risk will strongly influence the packaging of the project and the timing of tendering and awarding contracts. It makes sense to plan the contract packaging so that contractors are able to undertake work that is aligned with their particular skills and capabilities, whilst not requiring them unnecessarily to take responsibility for workscope they are not particularly suited to. Similarly, it makes sense to not ask organisations to manage and take responsibility for the performance of others if they do not have such experience or understanding.

A hierarchy for contingency

Where there is uncertainty, there is a need for contingency. This can be for both cost and time.

There is a tendency for clients to believe that contingency is something that should not need to be spent. Or a view that this contingency is only for the client to fund changes to his requirements.

There is a belief that the cost quoted is an absolute figure, and similarly that the programme is fixed. But on reflection with an informed client, it can readily be understood that any cost figure quoted has an associated probability of being the correct out-turn figure. Even the 'most likely' figure has by definition a probability of being exceeded. Budgeting just for this most likely figure is quite likely to be insufficient. The same logic can be applied to the most likely programme duration.

So the challenge is not whether there is a need for a contingency but how best to allocate it and how to manage and control its drawdown.

Contingency allowances are needed to cover the following:

- Resolution of incomplete information, as by definition in developing the complete details there are likely to be elements that were not foreseen.
- Overcoming circumstances that were not expected.
- Incomplete knowledge of the current asset condition or site characteristics.
- Correcting errors or mistakes.
- Needing to apply acceleration measures to avoid programme slippage.
- Covering the time-dependent costs associated with programme overrun.
- Allowance for the introduction of recognised nominal extra requirements or scope growth.
- Currency exchange rate movements.
- Indexing for inflation.
- Changes to legislation or statutory powers affecting the project.

Projects are likely to be more successful if the client has considered these and decided up front how to fairly manage the associated risks.

Risks need to be covered

When risks are placed upon an organisation that are significant, or with dependencies upon factors outside of the control of that organisation, then either there will be no organisation willing to accept these risks, so none prepared to bid, or they will put a high cost allowance into their price, which the client will have to bear whether or not the risk materialises.

So the client is left in a position where the risk is being uncomfortably handled by an organisation that is not best suited to mitigate it and is accepting a cost that is likely to be unnecessarily high. The reality is that the risk has not gone away, so the project will still suffer the consequences if it occurs, even though the client has attempted contractually to off-load it.

This is likely to lead to confused expectations, conflict and adversarial positions.

A more enlightened approach is for the client to hold on to those risks that he has most ability to manage and the most influence to resolve and to directly hold a contingency allowance for these. Then by clearly assigning other risks to parties best able to manage them the client can enter into open dialogue with each to reach agreement on what level of contingency these organisations should reasonably include in their price. It may be that they are also required to report on the drawdown of their contingency as a part of their reporting under the contract as the project proceeds.

A further option is to use a peer review process to assist in the quantification and management of risks. An independent group of experts can bring impartiality and past experience to bear and help reach conclusions on how best to assign risk.

This more transparent approach can motivate all involved to take a more proactive and considered approach.

Risks management

Clients should work to understand risks, on how to handle risks (including their impact on relationships) and their attitude and appetite for them before then considering allocation, management and risk registers. Clients' insurers will have a role also in this aspect of a project.

Clients should operate a 'live' risk management process, which in larger organisations is facilitated by risk analysts, who can assess ranges of confidence through quantified risk analysis (QRA) to help identify the likely contingency requirement. A project risk register should be first considered when formulating the business case at an early stage and then developed and used throughout the life of the project. It is widely accepted that effective risk management assists a business to achieve its objectives by

- Reducing the likelihood or consequence of negative events
- Identifying opportunities that would have a positive consequence
- Identifying and understanding complex multiple cross-organisational risks
- Supporting cost control and programme financial governance
- Developing best value through optioneering
- Providing visible and auditable governance across all levels of an organisation
- Protecting reputation and stakeholder confidence
- Increasing out-turn confidence for funders and insurers

In looking at risk it is useful to consider the issues of 'cause' and 'consequence'. It may be that putting all the effort into the symptom is misplaced if the underlying cause is not addressed. For instance, some clients believe their supply chain is not performing effectively on their projects. They will put great effort into rigorous processes to examine the effectiveness of each member of their supply chain to deliver. They may even set up departments with a head of supply chain management, with an importance and remit to create processes and run audits. However, these are often cases where they have appointed and are selecting firms with excellent proven track records from many other projects. What these clients need to consider is whether this poor supply chain performance is the cause or the consequence of their difficulties. Risk management should not only

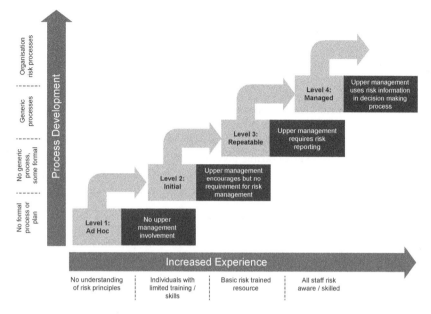

Figure 11.1   Progressing towards mature risk management.

be looking at others. It might be that the biggest risks are endemic within their own organisation and are a consequence of the client's structure and approach. If the client is setting up and in effect generating an environment of uncertainty then the poor supply chain performance will tend to be a consequence.

> *Doing the same thing over and over again and expecting a different result has famously been described as a definition of insanity.*

Clients strive to develop their processes and experience in risk management. The risk management process should be:

- Systematic
- Suitable for, and across, all levels of an organisation and project team
- Efficient
- Consistent
- Managed actively for the life of the project
- With a feedback loop to avoid repeating the same failings

Risks hierarchy

Avoid making risks too complicated, with risk registers that are so voluminous that those risks needing to be seriously addressed are lost amongst many others that are relatively unimportant.

Experience has shown a relatively effective filter is to use the concepts of 'impact' and 'probability', where

Impact = the degree to which a risk event would detrimentally influence the project out-turn

Probability = the likelihood of this risk event materialising on the project

Using a simple qualitative assessment of each of these on a scale of 1–4 and then multiplying to get a 'severity rating' gives a quick visibility of those risks to focus on. Typically those of severity greater or equal to 9 merit senior management attention, while those of lower severity can be left at the working team level.

Project Key Risks				
Risk No	Risk	Impact	Prob	Severity
6939	**Interface Management** – risk of ineffective management of key interfaces (ie, third party, contractor, local authority etc) and critical effects on construction programme.	4	4	16
6796	**Rail Contractors** – performance in Staging Works (including appropriate contractor resourcing). Risk of further knock-on delays to critical works.	4	4	16
6794	**Thameslink Box Blockade** – risk of failure to complete works withing blockade period.	4	3	12
6798	**Interim Station** – station operator requirements need to be finalised. Risk of non-acceptance and inability to open on schedule due to third parties, including HMRI.	4	3	12
5806	**Immunisation/incompatibilty of electrical systems** – risk of incompatibilty and interference between electrical systems/circuits and electrolytic corrosion to connected systems/structures.	3	3	9
6947	**Northern Line** – risk of damage to existing LUL tunnels from piling.	3	3	9
6811	**St Pancras Station** – risk of not finalising station operator requirements (including retail fitout) in a time	3	3	9
6399	Unsigned agreemer with EWS and Trans			

SEVERITY RATING								
		Catastrophic 4		Critical 3		Serious 2		Marginal 1
Very Likely	4	16	Intolerable	12	Intolerable	8	Intolerable	4	Moderate
Likely	3	12	Intolerable	9	Intolerable	6	Substantial	3	Moderate
Unlikely	2	8	Substantial	6	Substantial	4	Moderate	2	Tolerable
Highly unlikely	1	4	Moderate	3	Moderate	2	Tolerable	1	Tolerable

Figure 11.2   Managing Risk more easily using a Severity Rating Matrix.

Systems engineering approach

Some clients are increasingly looking to 'systems engineering' to deliver their projects. This approach considers both the business and technical needs of clients, with the aim of providing a quality system that meets their goals.

The systems engineering approach is comprehensive and structured. It enables project teams to develop appropriate integrated solutions that can be implemented as processes, products or services supported throughout the project lifecycle to the satisfaction of stakeholders.

Systems engineering should encapsulate the following aspects of a project:

- Systems engineering management (delivery strategy)
- Systems integration
- Requirements management
- Interface management
- Logic-linked programme with critical path
- System architectural modelling
- Value engineering
- Demonstration and testing
- Verification and validation

Clients should put a business management system in place as early as possible as part of the systems engineering approach. This will assist greatly with

- Project process development
- Document management
- Project configuration management
- Asset data management
- Asset knowledge management
- Asset condition monitoring and systems operation
- Knowledge management

If implemented at the outset and utilised throughout, this approach will assist in delivering the project objectives. However, it is important to ensure that a systems engineering approach does not get overcomplicated to the point where getting the systems to work becomes a project in itself.

The processes established need to be flexible enough to cope with the demands of the project, and also be fully understood by both client and supply chain. There is no point in putting processes in place unless people understand them, and if they are not understood this will undermine the whole project. A system that is all information and no knowledge may keep lots of people occupied but is of no use to the senior project management.

Programme management

For many clients their biggest risk is adherence to programme. So programme management is an important tool for clients – 'time is money' – and is under constant scrutiny. Experienced clients therefore ensure the programme is properly developed at the outset and in doing so make some risk-based assessment for programme contingency or float.

Change is almost inevitable in any project and can be imposed from either internal or external sources. Clients need to be able to deal with this and have change control procedures in place as part of their systems engineering architecture, so that the effects (e.g. delay or acceleration) on the critical path of their project programme can be picked up immediately. Such controls encourage a 'no surprises' management approach and enable any necessary changes or workarounds to be implemented as soon as a change is identified. These will ensure that all implications of the change are addressed – cost, programme and interfaces.

Programme management can be controlled by reporting progress using SMART objectives or targets:

S = specific
M = measurable
A = achievable
R = relevant
T = time-bound

Controlling costs

Many factors have an influence on cost. Clients should set up their projects with factors that will make costs more controllable and give more certainty on out-turn. The factors include:

- Defined scope, with the project 'boundaries' established
- Clear objectives
- Whole-life costing
- Realistic budget, with adequate risk and contingency pots
- Achievable programme
- Incentivisation
- Systems engineering approach
- Change control
- Risk management interface management
- Clearly defined forms of contract, with parties understanding what they have to price for

Budgets have to be allocated fairly and monitored regularly, together with risk and contingency pots.

> *Things got off to a bad start when we did not have an adequate budget. We should have stopped things right there until we had got that sorted out. Instead we rushed into it. On most of our schemes now there would be a risk register that would have ensured before we started that it had an adequate budget. But on this scheme there was very little done to ensure that.*

Many clients use a quantified risk assessment approach to setting up the project budget, in which levels of confidence relating to cost and programme are evaluated and appropriate contingency allowances included.

Establishing the project cost and programme

The challenge for both cost and programme is that there are no absolutes. There is no single 'right answer' that can be used at the beginning of a project and relied upon to be correct at the project out-turn. There are terms used like the 'point estimate' for the cost plan, but the contexts for these need to be understood before they are accepted. The term is referring to the percentile point of confidence.

A good approach, therefore, is to break the project into its major component parts and produce a cost and programme estimate for each, together with a range of confidence. So an expected figure, together with a best-case optimistic figure and a worse-case pessimistic figure. The range of this spread will tend to indicate how well conditioned that element of the project is and how much risk is associated with it. This can either be from the level of confidence in the definition of that component and its objectives, or in the associated prevailing conditions. This builds a 'three-point' estimate for each component.

A probabilistic review can then be run to assemble all the components of the project into a combined 'three-point' estimate. From this an informed risk based view can be formed of the expected out-turn cost.

Importantly, the same needs to be done for each component for the expected delivery timescales. One of the biggest influences on out-turn cost is time, so this programme confidence needs to be fed into the three-point cost model. Almost always there are significant project costs or establishment overheads that are determined by time taken.

Other important factors in reaching an overall view on confidence in out-turn cost and programme are the effort required for up-front approvals, planning and consenting phases and then the commissioning and operational-readiness phases.

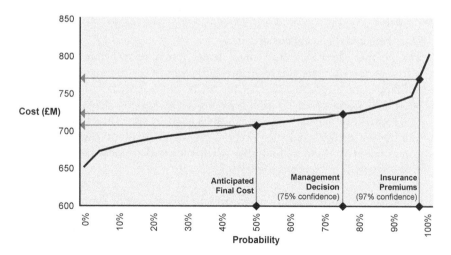

Figure 11.3   Probabilistic Out-turn Cost model – to illustrate distinguishing between 'most likely' and 'management decision' values.

Having put all of this together it is possible to represent it for both cost and programme and take a management view on whether the project is well enough conditioned to be able to proceed with confidence. The mid-range zone of the model should be reasonably 'flat' to be well conditioned. The 'percentile points' selected for management and other purposes can then be selected based on a clear understanding of the underlying issues and appetite for risk.

Some popular misconceptions in controlling risk

For many clients, four topics are often associated with controlling risk:

- The design process is difficult to manage
- Early contractor involvement is perceived to be beneficial, but they are not sure why
- Eliminating interfaces is desirable
- Value engineering can come to the rescue

When a project is completed '*on time and on budget*' the project manager and the contractor are often praised for doing a great job. But when the project overruns or is too expensive it can be the designer who is blamed, because the design was late or unbuildable or over-specified. The perception is often that the design was not controlled, and a common retort to perceived inadequate design management is to involve the contractor early to manage the design. This is additionally seen as beneficially removing the interface between design and construction.

However:

- Contractors' core skills are in building, not designing
- Contractors require their best staff to be leading large teams of their own staff, managing and directing construction
- Placing these best staff into client teams, often years before construction starts, means they are not available to do what they are good at, which is building
- Most UK contractors do not have in-house design capability, so early involvement to manage the design results in them subcontracting the design
- Contractors will give one view on pricing when there is not a firm contractual commitment to them and another when they understand what commercial risks they will be expected to pick up

Going with early contractor involvement and leadership can therefore result in:

- The client often getting a design team of the contractor's choosing and aligned to the contractor, rather than of their own choosing with first allegiance to the client

- The contractor being positioned to determine what the various aspects of the project cost to suit his risk profile, rather than their true value to the client
- The client being committed to a contractor before the project has been adequately defined and before decisions have been made on how best to assign risk, ownership of programme/schedule float and cost contingency

So this may not really be the right answer. It may be better to have client control of the design development before making commitments to the major costs of construction. Unnecessary interfaces are definitely worth avoiding, but it may not be beneficial to give up interfaces if this also results in losing a point of control.

However, the design needs to be buildable and deliverable, so, on balance, the need may not be for early contractor involvement but for early construction planning skills which enable the development of a procurement strategy and an understanding of how complex phasing and construction interface issues should influence the final design.

Understanding design

At the inception of the design process there is a wide range of possible options by which the end result can be achieved. At the end of the project there is a single design solution – the one that is built. Designers will need to produce information through the various stages of the project, to progressively translate the client's brief and objectives into information for construction.

The early stages are the opportunity for experienced designers to identify different ways to achieve the brief and to bring their creativity to illustrate, through dialogue with the client, the optimum manifestation of the client's ambitions. This is a key defining stage of the project and one in which clients should be actively engaged. It results in a preliminary or concept design that should be accompanied by an indicative cost estimate and advice on constructability, to underpin that the design concept is affordable and viable.

The successive stages are to then develop this preliminary design concept so that it can be competitively priced, used as the basis for construction contract award, information for planning and procurement (IPP) of the works elements, information for construction (IFC), manufacturers' shop drawings, through to the as-built record. The control of these stages is fundamental to the success of a complex project.

In considering design progress, each stage results in greater definition, and in consequence a progressive reduction in the range of options. The concept of a 'design freeze' is often misunderstood. It is not possible to freeze design absolutely as it is necessary to progressively develop and finalise the level of detail, but the range of influence for this design development must become progressively more contained.

Figure 11.4   Design Process – through controlled Design Stages.

If the design is under control it is in an acceptable range of influence within the design envelope appropriate for the stage reached. If ideas are being considered with a broader zone of influence than is appropriate, then the design process is not under control, action is needed – either to bring it back to within the planned constraints, or to accept change and an extended completion date.

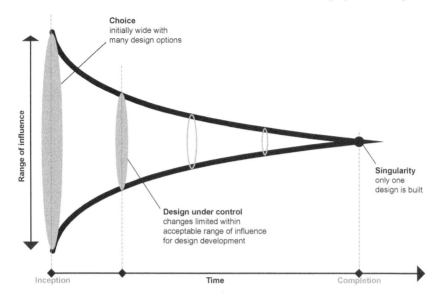

Figure 11.5a Managing the design envelope of uncertainty – 'under control'.

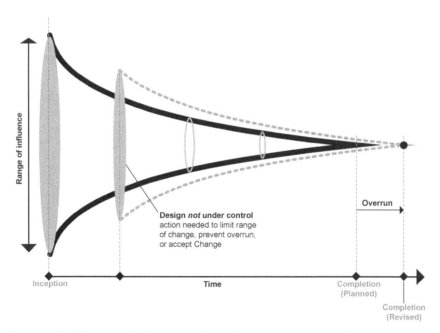

Figure 11.5b Managing the design envelope of uncertainty – not 'under control'.

Managing the design process well not only requires an understanding of the level of detail expected as the design progresses through its various stages, but also an expertise in the integration and coordination of the various design disciplines. It is recommended that a principal design consultant is appointed to fulfil this lead responsibility. This is an additional duty over and above the particular design disciplines provided by this consultant. For building projects, the architect commonly took this principal consultant role, but as projects have become more technically complex there has been a trend away from an architectural lead and a move towards the lead consultant role being fulfilled by an engineering consultant with multidisciplinary design capabilities. Major infrastructure or civil engineering projects, more traditionally, are engineer-led.

Value engineering comes with risk if introduced as late design change

Value engineering is often misunderstood and misused, with clients unwittingly inviting changes that encourage a loss of control. An example of this is clients encouraging value engineering after contract award because they need to make savings, or by acceptance of alternative bids as cost savings. These are an all too easy way of permitting change outside the control design envelope.

Why do contractors, once appointed, welcome value engineering?

- The contractors can appear to be to be protecting the client's interest, introducing options for savings, with relatively little substantive input or background knowledge, in a non-competitive environment.
- In doing so the contractor can reduce their risk by de-specifying the project to a greater proportion than the associated savings they offer.
- As a consequence, the contractor introduces late changes to the design, which they are often not responsible for implementing and which tend to undermine the design team's position in leading the design solution.
- These changes have to be worked through and incorporated by the design team, which puts the design team under new pressure and is likely to result in late release of construction information to the contractor and thus scope for claim for delay by the contractor.

So through value engineering the contractor often gets an increased return margin to reduce the performance of the project and increase the programme. It is not at all clear that the client gets real value for money.

Designs must be able to be commissioned and maintained

The success of complex projects often relates to their ability to be put into operation and efficiently maintained. This is surprisingly often overlooked or not given sufficient attention by the right people at the right stage in the design development. Early input, if recommended into the design process by those who will be responsible for commissioning and maintenance, so that the appropriate

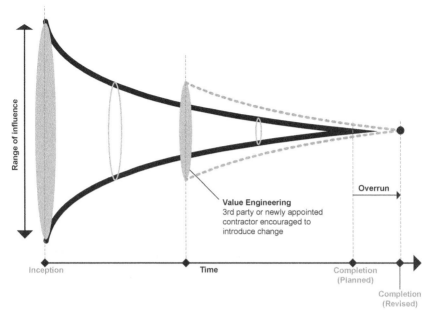

Figure 11.5c Managing the design envelope of uncertainty – 'Value engineering too late'.

provisions are included into the design. In the case of projects that fit within a greater campus or network it will be important to establish the control interfaces between the project and the site-wide systems and infrastructure so that the project commissioning fits within this greater context.

Out-turn cost – anticipating risks in design development and construction

The best projects result from designers that are creative. Design development is a normal process, but it does not come free, and often clients are surprised by this. The design envelope starts with an initial three-point estimate of out-turn cost – an initial point estimate and an upper and lower cost-certainty range. Given this range, it would be naive to assume that with all the design development the out-turn cost figure would remain absolutely horizontal at the initial point estimate on the graph.

As the level of design information is developed, and increased understanding is gained of particular requirements, it is normal for the out-turn costs to be above the initial point estimate.

Similarly, as the project progresses into construction, unforeseen difficulties and unexpected site circumstances are typically encountered. It is expected that the contractor will find effective ways to overcome these but they will also put upward pressure on the out-turn cost.

So it is necessary to establish a project budget that allows for a contingency cost above the initial point estimate. It will be this project budget, and not the initial point estimate, that is the reference for assessing if the project comes in within cost.

But this is not a static position. The range of cost uncertainty also changes as the design and construction progresses, normally reducing as more precise information becomes available or work is concluded. So the median figure (updated point estimate) may be increasing but the range of uncertainty is converging. This requires frequent cost evaluation as the project progresses and an awareness of the range of risk or uncertainty relative to the project budget and the initial contingency allowance.

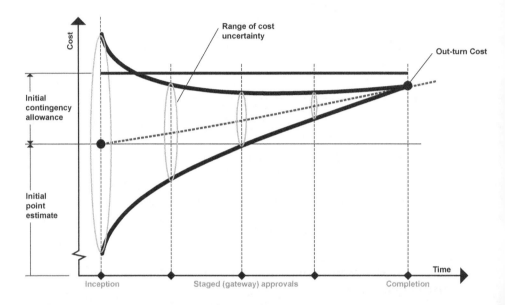

Figure 11.6a   Managing Cost Certainty – towards an Out-turn Cost.

The challenge at the beginning of the project is to assess what the appropriate contingency allowance should be. It cannot be pitched so high as to cover extremes that are not required. If too excessive it is likely to be too high a hurdle for the project to gain financial close and approval to proceed. If too low then the available overall budget allowance will be exceeded and the project will not be delivered within budget.

The effective drawdown of a contingency is a crucial aspect of the proper management of major projects, and the maintaining of headroom between the budget, including contingency relative to the projected actual out-turn cost. Often this is not clearly communicated to clients, so inadequate understanding of cost imposes unrealistic constraints on the design development and an unnecessary adversarial approach when in construction.

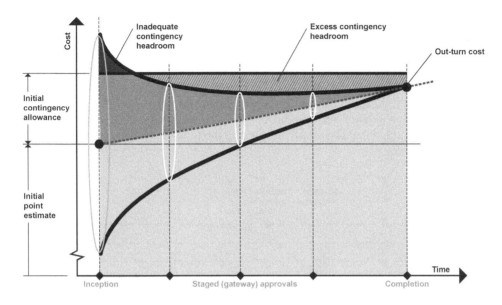

Figure 11.6b Managing Cost Certainty – the consideration of Contingency.

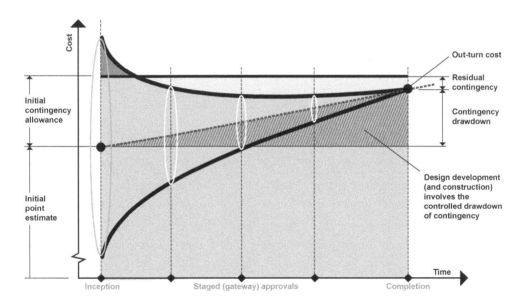

Figure 11.6c Managing Cost Certainty – monitoring Contingency Drawdown.

Out-turn programme – anticipating the likelihood of the completion date

The project programme, like project costs, comes with a range of certainty. The most likely completion date, by definition, is not a good basis for planning with confidence, since there is probably a 50 per cent chance of a later date being actually achieved. It therefore makes sense as a client to plan with a higher than 50 per cent confidence and use this to set the expected 'on time' completion date. In turn this should be used when feeding the time-dependent costs into the out-turn cost model.

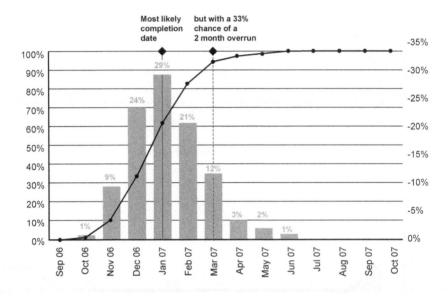

Figure 11.7   The range of certainty in the Completion Date.

12 Performance management
How to measure and control progress

Introduction

An effective performance management system fulfils two critical functions. First, it is used to translate the strategic intent into clear and meaningful terms to assist leaders communicate the strategy to everyone involved in the project, or programme of projects. This could be said to be making sure everyone will be *'doing the right things'*. Whilst this is crucial during the early development stage to achieve wide understanding, commitment and motivation, it should be continued throughout all stages of the programme to keep everyone familiar with changes that emerge from ongoing strategic and operational reviews.

Second, it provides feedback about progress being achieved towards the business goals. This flow of information enables decision-makers to continually appraise the project based on fact and data to ensure its success. It also enables the original strategic assumptions within the business case to be tested and adjusted in the light of emerging issues. This could be said to be making sure everyone is *'doing things right'*.

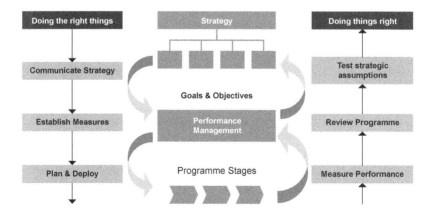

Figure 12.1   Strategy – from 'doing the right things' to 'doing things right'.

Managing expectations and performance

Performance management is a systematic approach to help manage and lead the delivery of projects, linking business goals developed in the planning stage to overall project outcomes. The approach ensures that project success is clearly defined, is measurable and is ultimately achieved. Leaders in best-practice organisations are personally involved in establishing the system, typically during the development stage, and remain actively involved throughout all subsequent stages of the project.

There are two critical roles for a performance-management system:

1. *Sharing of results:*
   Effective internal and external communications about project performance and its benefits is vital to successful delivery. On successfully managed projects – *'The clarity of the business plan was crucial; everything was clear and regularly communicated.'*

   This approach builds commitment, motivation and confidence amongst stakeholders. The performance management system is an important means to assist with this by providing sound information and data. A wide variety of communication channels should be used to get information out to people. While protecting sensitive financial information, best-practice organisations distribute their performance management information on their internet and intranet sites for real-time access by various levels of management, teams and individuals. They also use periodic reports, newsletters, electronic broadcasts and other visual media to set out their objectives and achievements.

2. *Accountability for results:*
   The performance-management system should be used to set out account-abilities for results that must be clearly assigned and well understood. To be effective, accountabilities should be agreed formally and signed off by all parties involved in their development to avoid misunderstandings and disappointment. To have any effective meaning at all, accountabilities must be associated with both positive and negative consequences but must not be punitive. Performance management should be positive and proactive, and not be the basis to apportion blame. Leading organisations focus performance management on improving the organisation and its systems. Individual culpability is only considered as part of a systemwide analysis and even then the focus remains on learning and development. In summary – *If you put good people in bad systems you get bad results*.

The behaviour of individuals is driven by the consequences of their actions, and so a fair, balanced and motivational environment should be developed in terms of compensation, rewards and recognition. It is one of the most powerful levers senior managers have to directly influence positive project outcomes because of its effect on behaviours. Organisations frequently underestimate the profound effect that this can have on overall performance – *It is always positive to work with a performance management system, and it absolutely does affect behaviours*.

Best-practice organisations use a balanced set of performance measures, commonly referred to as a 'balanced scorecard' approach. By focusing on prime project outcomes, as well as on what influences these outcomes – such as operational effectiveness and development and learning – these organisations build a more comprehensive view of their project programmes, which in turn helps them to act in the best long-term interests of all parties.

Performance measurement should provide actionable information to decision-makers and should be presented with the results of comprehensive analysis carried out by appropriate subject experts. The analysis should look for deviations from expectations and try to account for these, assess opportunities and risks, forecast future performance and finally make informed recommendations for change. Measurement becomes counterproductive if used to assign blame or to simply accumulate data and adhere with reporting requirements. The system should allow for control to be cascaded and managed at the lowest effective level.

Key performance indicators (KPIs) are commonly used on successful projects, with clients defining these up front and having them included within the delivery contracts in a consistent manner, so that progress can be similarly reviewed across different packages.

KPIs that are normalised are particularly powerful in providing a quick overview for the senior project-leadership team and client. Examples include:

- Cost performance indicator: $CPI = \dfrac{BCWP}{ACWP}$

- Schedule performance indicator: $SPI = \dfrac{BCWP}{BCWS}$

In each case a value greater than 1.0 is good news and a value less than 1.0 is in need of attention.

> Where BCWP = Budgeted cost of work performed
> ACWP = Actual cost of work performed
> BCWS = Budgeted cost of work scheduled

It needs to be remembered that '*what you measure is what you get*', so a rounded set of indicators is important to provide a balanced view. Though avoid setting so many indicators that they become a costly self-fulfilling industry in their own right, generating much information but little knowledge ('wood from trees') – *Critical to get KPIs aligned with strategic corporate objectives*.

Research has found that project leaders should take particular care to ensure that supplier organisations are capable of collecting and providing reliable data for measurement. There is no point setting KPIs that they do not have the means to measure and track.

However, it should also be recognised that many indicators to measure progress are backward-looking, assessing what has been achieved. This may be fine as long as the baseline against which the indicators are measured has credibility and as long as changes and variations have been properly incorporated. Where this is not straightforward, the measured progress and earned value approach can be misleading. So it is recommended to use a mixture of *'earned value'* analysis and *'forecast to complete'* assessments. If the two do not compute to give a reasonable fit against the project timescales and budget then something is not under control and needs detailed evaluation and corrective actions.

Measuring design progress – 'earned value' or 'forecast to complete'

It is common to plot the cumulative man-hours spent on design as an 'S' curve, which implies a Gaussian distribution in terms of resource effort over the course of the project. However, from experience, the last 5 per cent of any design delivery takes 15 to 20 per cent of the effort, giving the more realistic graph in the illustration below. The standard prognosis is shown as a dotted line and the more realistic outcome is represented as a solid line. This should be planned and allowed for.

If the man-hour spend is lower than planned at a point in the design there can be the impression that this is good news. But this is meaningless without also an assessment of progress achieved. If behind, then a recovery plan is needed and a revised view of the likely completion date. In any programme some activities are time dependent and others are deliverable dependent. If you are behind on deliverable dependencies you may not be able to recover your time-dependent activities in terms of resource.

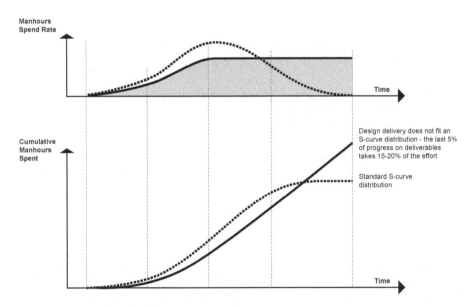

Figure 12.2   Measurement of spend on design progress.

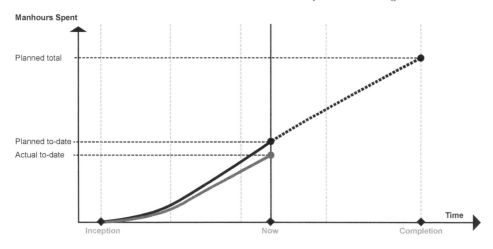

Figure 12.3a Measurement of design progress achieved.

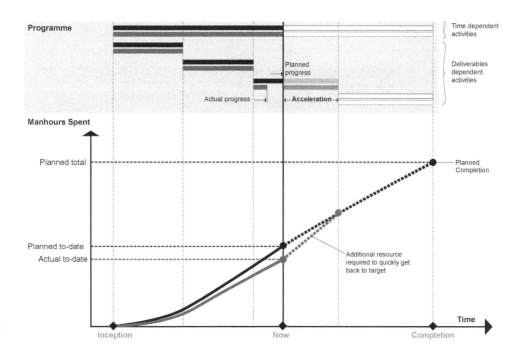

Figure 12.3b Measurement of design progress and earned value.

In many respects, when assessing design, 'forecast to complete' analysis is much more useful than 'earned-value' analysis. Earned-value analysis has recently become popular with clients and may tell you where you are, but what you really need to know at any given point is what you need to do to get to completion. The series of diagrams in Figure 12.3 is a representation of component activities and the degree of completeness of the deliverables associated with each. The logic is to assess and forecast levels of input required against the remaining planned activities. Again, some activities can be correlated with deliverables, but others, such as team management and design co-ordination, tend to be related to time rather than deliverables. These time-dependent activities need to be understood in preparing forecasts to complete.

A further reason for forecasting to completion being much more relevant is that earned-value analysis can be misleading due to the common tendency for creep in the number of deliverables relative to the initial plan, and thus a false sense of completeness when making earned-value assessments.

One common mistake is to measure deliverables at too fine a scale. For instance, attempting to measure by counting individual drawings rather than the design element that might be represented by a number of drawings. Creep in number of deliverables can easily occur by adding in extra drawings, whereas the element is less prone to quantity variation when measuring progress.

So remember that design is a non-linear process, and that the last push to get fully complete documents takes a greater effort than usually anticipated, and that there is a tendency for upward creep in the number of deliverables.

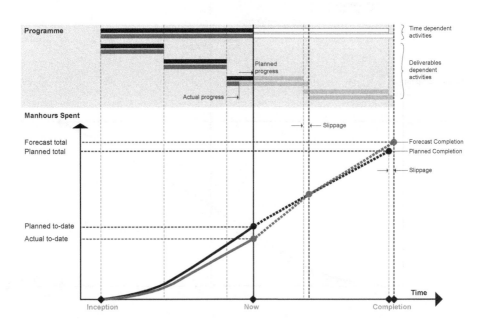

Figure 12.3c   Measurement of design progress and Forecast to Complete.

Transition from design to construction lead

There are different skills required for design leadership and construction leadership and these tend to attract different types of people. The consequence is that it is good to achieve some form of controlled and planned transition between the two, as both are clearly needed. An abrupt interface may lead to a polarisation of respective positions. It is better to enable a constructive dialogue so that an understanding can be built between the two.

Good experience suggests that the package tender process is a sensible way to plan this transfer, with the design lead pushing the package into the process and the construction lead pulling it out. Achieving a shared involvement of both through this, with them working jointly. Having the emphasis progressively switch means they have a shared interest in getting to the right result.

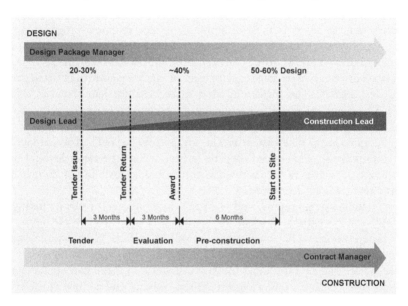

Figure 12.4 Transition from design to construction lead.

Measuring construction progress and spend

There has been a tendency for evermore complex logic-linked programming to be used for the assessment of construction progress. This is in the expectation that if everything is fully planned out, then confidence in the final deadline can be monitored and maintained. Critical paths can also be identified.

However, on complex projects this approach is often incredibly complicated and there can be multiple critical paths depending on allowances for float (the time between the completion of one activity and the commencement of a follow-on activity). Such programming can be very time consuming, expensive, and still being updated or corrected when management decisions are needed; so not actually helpful for the high-level monitoring of the project. Variations, in turn, result in a major effort to incorporate, and the original baseline control position can be lost. Quite fast this logic-linked programming can result in 'not seeing the wood for the trees, if not even attempting to count branches and leaves.'

> *The project programming was well intentioned but resulted in 'all information and no knowledge'.*

The programme time and work achieved to date is one of the most important aspects of measuring construction progress and in controlling the out-turn cost.

> *Keep in good control of time and there is only so much you can spend.*

So, for the client, it is the big-picture understanding that is important. An effective way to see this is to assemble a project-overview programme. This identifies the major component packages and key milestones within each, with particular emphasis on identifying those milestones that link as a primary interface between one package and another. If you keep control of these primary interfaces, then it is less important what happens at the detail level in between them. The client and the senior project management focus on the primary interfaces and any corrective actions needed if there is a likelihood of one of them slipping, or scope to take benefit if one can be achieved early. The management of the individual packages can then be left to manage the lower level of detail. This hierarchy achieves the required visibility and accessibility of knowledge to inform management decisions.

Very detailed programming and even 3D visualisations (or 4D, including cost loading) can be extremely useful for particularly critical work elements, such as railway possessions or power plant outages to include 'go' or 'no-go' options linked to programmed interim checkpoints in order to ensure handback on time. These are specific circumstances in which the level of detail can be manageable and can really enhance understanding of the viability and control of the critical construction activities.

It needs to be realised that contractor's costs are divided between

- Time-dependent fixed overheads (e.g. site establishment, security, management)
- Work executed
- Plant and equipment costs
- Fees

Of these, it is the work executed that is resulting in the project that the client wants and the progressive value added in moving towards its completion. So the incurred proportion of the overall planned spend may not be a useful indicator to reflect actual progress achieved.

It is worth tracking these different contractor costs separately in order to gauge the effectiveness of the construction performance. On large, complex projects with multiple packages it is worth being very clear on where responsibilities for the site establishment lie and how these are transferred at key interfaces to avoid double-costing. Losing control of time can result in these fixed costs becoming very dominant, and are a major reason for project cost overrun.

Some forms of contract reimburse the costs of plant and equipment, in which case keeping a close track of equipment utilisation is important in assessing construction performance.

The essence of reviewing construction performance is to have usable information to enable collaborative and constructive decisions to be made to help keep the project on track – a *'no surprises' approach*.

> *For contractors, a form of 'dashboard' report was developed. When the project started, contractors' reports were very thick with all kinds of information buried in them. This meant that if something was contested, the contractor could say that it had already been reported and we had failed to pick it up. So it was safer to have just one A3 sheet of paper laid out to report and monitor facts on the key activities. Wherever possible, graphs were used so that trends could be reviewed. This ensured everyone was aware of the problems and made it easier to agree and resolve the issues at monthly meetings.*

Safety critical control – but avoiding unnecessary constraints

In many industry sectors (a couple of examples being rail and nuclear) construction activities take place in or adjacent to 'live' operational, safety-critical facilities. This needs careful recognition and planning because there is often a significant cost and programme impact of being within the operational controlled zone.

There is a need to consider activities under three categories:

- *'Green field' working* – site area outside of the operational or controlled secure zone, where normal construction with a workforce that does not need specialist training in the risks of the controlled zone nor the constraints associated with the rigour of operating within this zone. This construction can be much more flexible, availability of labour can be greater and in consequence the costs can be considerably lower.
- *Controlled operational zone working* – areas requiring the full security of access and special procedures and training, such that only a pre-accredited workforce can be deployed, with work constrained by the absolute need to comply and demonstrate compliance with safety-critical industry-specific standards.
- *Possession working or Outage working* – this is a further level of criticality in that it is activities within the controlled operational zone that are such that the normal operational activities have to be temporarily suspended but able to be put back into use in a very controlled way and with a high level of confidence.

In order to plan work most cost-effectively it is very important to understand which of these categories apply, and wherever possible to ensure work is not taking place in a higher category than is really needed. There may be a need to challenge the attitude of, 'We have always done it like this' in some industry sectors. For instance, a significant piece of work may not really be a temporary possession and may justify transferring a whole site area out of operational status back to a 'green field' working status.

A further factor to consider is who or what is being protected through these different categories. Is it:

- The operational process and those involved in operating it?
- The construction workforce needing to be protected from exposure to operational risk?
- The public who may either be using the facilities or reliant on them?

The investment justified and the levels of security and management-control measures put in place through construction need to reflect these issues. The consequences of getting this wrong are significant – either not having the necessary safety-critical controls in place, or of spending unnecessarily on a level of constraint that is not needed.

High-level project overview programme and key interfaces

An informed client expects to know what the high-level key milestones are in the next month and whether these are going to be achieved. If those in the previous period were successfully achieved and the expectations are good for the coming month then the likelihood is that the project is reasonably under control.

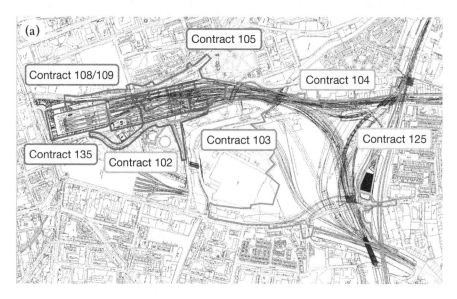

Figure 12.5   An example of critical interface dates on a multiple package project – monitoring the key interface dates and avoiding too much information concealing the big issues.

(b)

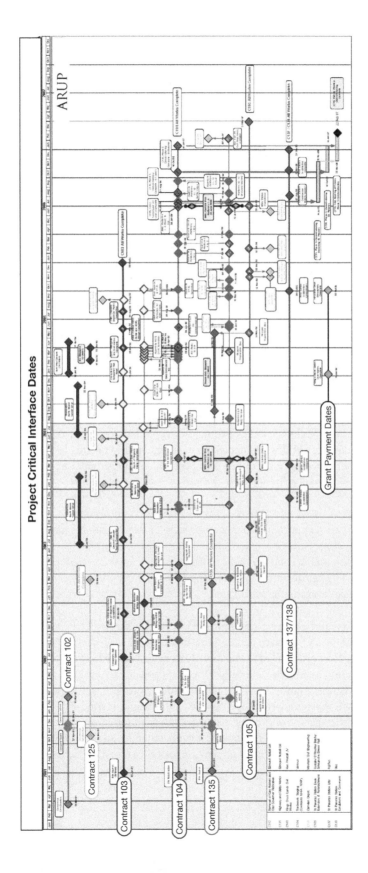

Project Critical Interface Dates

ARUP

Milestone events usually relate to interfaces. Either between separate activities or packages within the project or in interfaces with third parties. Therefore a good high-level programme will flag up these key interfaces as milestone dates.

If the key milestone dates are not being achieved, or if circumstances have changed to put a milestone at risk, then there needs to be constructive intervention at senior level to establish a workaround or a recovery plan. If neither of these is possible then there needs to be a resetting of the programme and communication of the consequences to the affected packages and third parties.

> *Understand and keep control of the key interface milestones and you have the basis for successful control over the project programme.*

Execution plans and dashboard reports

Having a plan but not having many people know about it is not of great use. Equally, having reports on progress but not being sure what they are benchmarked against is less than ideal. So having a high-level execution plan that summarises the reference baseline is useful. However, continually updating and changing this reference baseline can be an excuse for loose control and management. Often there is plenty of worthy but distracting activity in redoing the plan, rather than in making the previous plan work. If everyone knows it will be updated again in due course, then why try too hard either to get the revised one really right or to communicate it. There is a danger in all of this that there ceases to be a single version of events, or even a lack of clarity over when things actually occurred and when decisions were effectively taken.

Another aspect of reporting is to report on so much that the really important information gets lost in amongst the rest.

> 'When I asked why a really critical issue had been overlooked, I got told no it hadn't. It was in clause 15.3.2 on page 38 of last month's 69 page progress report and it clearly said that some decision was needed. But given its obvious importance, which has now come seriously to light, I would have expected it to have been on page 1! The frustration is that I can't now claim for the cost of the consequences, nor take measures to avoid them.'

The clarity and effective communication of the big issues is what matters for senior leaders. A means to achieve this on complex projects has been to require a 'dashboard' report. Ahead of the monthly executive meeting all of the important details summarising the condition of the project, an awareness of the major progress against planned and the required decisions and sign-offs, to be included on a single A3 template. Accompanied by a single second A3 sheet behind it to summarise actual and earned spend against planned spend. For busy leaders on both sides this gives the means for an efficient and effective discussion at the monthly executive, and agreement to be reached on where the required actions are.

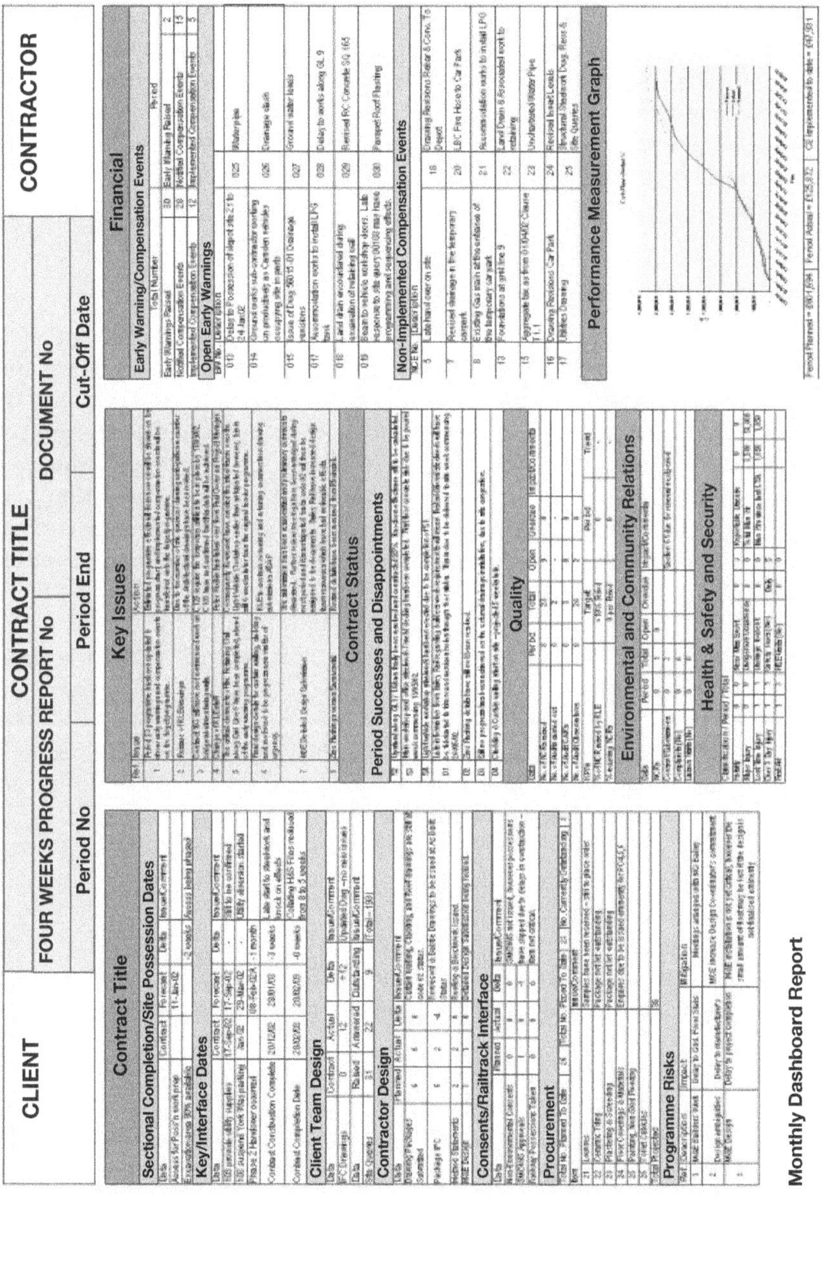

Monthly Dashboard Report

Figure 12.6 Dashboard report.

A good anecdote is that a German project manager was asked to go out to Brazil to lead a project to build a new engine factory. He was experienced in the motor industry but not in managing a construction project with multiple contractor inputs. Every month he would receive weighty reports submitted by all of the contractors and try to work out what they were telling him. All he wanted to know was whether the project was going well or not and, if not, what needed fixing. One evening when driving home he suddenly thought that the dashboard on his car was doing exactly that. It was telling him how fast the car was progressing, when it needed an input (of fuel) and by exception told him if anything needed attention so he could get it corrected. He went in the next morning and asked for dashboard reports.

Raising performance standards and performance review

Benchmarking should be used to establish performance targets as part of a continuous improvement process to ensure that the programme remains at the forefront of best practices, continually raising performance expectations. By systematically comparing key features of the project against other leading project organisations internally and externally from across the sector and wherever best practices are found, opportunities for improvement and innovation can be exploited for the overall benefit of the project. Best-practice organisations recognise that everything is relative to competition and best practice, so set up performance measures to reflect this.

A good client organisation gives decision-makers the delegated authority to make decisions in their defined remit.

Targets should be used to set challenging but achievable objectives throughout the programme, and should be developed and continuously reviewed on an informed basis, taking account of factors such as:

- Past performance – baseline data
- Competitive and best-practice benchmarks
- Stakeholder expectations
- Project strategic and operational needs, including future expectations
- Senior management and leadership ambitions for the project

Best-practice organisations systematically review programme performance at strategic and operational levels on a regular planned basis. This is considered to be the most important routine activity that programme managers and leaders undertake to secure the ongoing capability of their programme throughout its various stages. To ensure that the performance management system continues to provide relevant and timely information, they continually assess whether their current measures are sufficient or excessive, are proving to be useful in leading and managing the programme, and are driving towards the desired results. Targets should also be reviewed regularly, but too frequent changes in both measures and targets can cause confusion and affect accountabilities and motivation. Changes in performance criteria in the life of a project should therefore be kept to a minimum, but if required should be introduced systematically, progressively and with wide communication about the reasons for change.

13 Construction expertise

How to involve the best contractors and SMEs

Introduction

The implementation stage of any project is fundamentally important to a successful out-turn. This is the stage when the greatest expenditure occurs, so will determine whether the project is brought in under budget or not. Maintaining or beating the programme is also important during the implementation stage as there is a great tendency for the best way to control overall cost being to control time. It is also the stage when the desired quality is either achieved or is not.

The implementation stage centres on construction, but also involves pre-construction activities. Then, following construction, it involves activities associated with commissioning in order to hand over the project ready for operation.

Very few client organisations have in-house construction capabilities. It is normal to appoint contractors to carry out these activities. The best contractors have proven expertise in related areas and the types of construction involved in the project, have good connections with the construction industry supply chain and can bring strong buying power through these established links. Additionally, they can bring a strong safety culture and the proven ability to achieve high health-and-safety performance.

What does the client want the contractor for?

The choice of the right contractor may not be as obvious as first thought. There can be various expectations behind the appointment of the contractor:

- Is the contractor wanted for constructability advice ahead of the construction and, if so, is the same contractor then expected to go on and deliver?
- Is the contractor being selected based on the ability to actually build or the ability to manage others?
- Is there a need to appoint a principal contractor and does the client then expect this principal contractor to take direct responsibility for whichever subcontractors are needed?

- Is the contractor being appointed to provide a turnkey delivery and take over all aspects of implementation of the project, including any remaining design?
- Is the contractor being selected to bring international capability into the project because local contractors do not have these skills?
- Is the project large enough that it might require a joint venture of contractors in order to provide the combined capability required?

Many of these decisions influence the procurement strategy and the form of contractor procurement that should be selected. These were described in an earlier chapter. There is no sense in selecting a procurement strategy that is likely to result in a contractor being appointed without a good match to the client's expectations of the contracting organisations. Reviewing and understanding these expectations is an important step towards involving the best contractor for the desired objectives.

Early contractor involvement

The two reasons for involving the contractor early are:

- to obtain constructability input in parallel with the project development
- to require the contractor to manage the design

A number of contractors are now positioning themselves to be experts at providing early constructability advice. In effect they are doing this as an advisory service as construction solutions providers. There may be circumstances where they see this as the opportunity to provide creative input to projects on a fee basis when they would not want to carry the actual construction risk in the role of delivery contractor. Or where the project risk profile is such that the cost of competitively tendering for the construction role is too high to be worthwhile for the contractor. However, in the role of constructability advisor the contractor can provide real benefit to the project in checking that it is deliverable and in helping to plan the key construction sequencing. This knowledge of construction can also benefit the assessment of the environmental impact during construction – required working hours, material delivery transport routes, the size of the site establishment, labour requirements and maximum numbers on site, opportunities for recycling of demolition materials and spoil, temporary power requirements and the site security provisions.

If providing this early constructability input, there is the question of whether it is sensible to also allow this contractor to bid for the subsequent construction? Depending on the procurement route, it may be better to keep the roles of constructability advisor and construction delivery separate. This will avoid the potential danger that other contractors see the incumbent as in a preferential position and therefore these others are less inclined to compete in a tender process.

When appointed early to manage the design, the best contractors can demonstrate a real expertise in managing others, and in particular are able to manage those parts of the process for which they do not have in-house understanding as a part of their core business. If being asked to manage design they need to be able to demonstrate an understanding of the design process and the means to achieve sign-off and approvals. Are they able to plan and manage third-party independent design checks where these are needed? Do they understand planning approval processes and industry standards and compliance requirements? An experienced construction director may be excellent at managing activities on site but may not be best suited to managing pre-construction activities. Managing design may for many contractors be seen as a big risk, suggesting that they are not comfortable taking responsibility for it.

Some contractors, on the other hand, are excellent at taking an outline design and developing it to suit innovative and cost-effective construction methodologies. They will fine-tune the detail of the design concept in order to optimise temporary works and construction staging. This may even include establishing off-site prefabrication or setting up consolidation centres for just-in-time delivery to site. To make these work effectively, aspects of the design may need to be specifically developed so that prefabrication off-site is possible. The interfaces between different elements of the design may need to be adjusted in order to suit. Joints will be needed between the various prefabricated elements which might not have been needed if everything had been built in-situ. Prefabrication can take work off the critical path programme and thus tends to increase delivery confidence. It may suit a contractor's ambitions to be innovative and take risk out of the project, but it might not be a cheaper option. It might also influence, either for good or bad, the visual appearance of the completed works due to approaches to standardisation to benefit prefabrication. Elegant construction can look good but needs an associated expertise in giving attention to details. Prefabrication cannot therefore be imposed as an afterthought. If desirable, an experienced contractor needs to be involved early.

Individual contractors are likely to approach such prefabrication slightly differently, to suit their particular construction equipment, materials handling, transport and cranes, direct labour skills, approach to details and their preferences for supply chain trade contractors and supplier relationships. Advice on prefabrication provided by one contractor may therefore not be ideal for others.

When to transfer control to the contractor

Involving the contractor early can be beneficial, but there is then the debate over the point at which to have a formal contractually binding obligation. This tends to split between going towards a construction management route with trade contracts for individual packages, or going towards a construction D&B (design and build) route.

The first of these is likely to achieve the lowest overall cost but with the risk that this total cost will only progressively emerge as the trade packages are

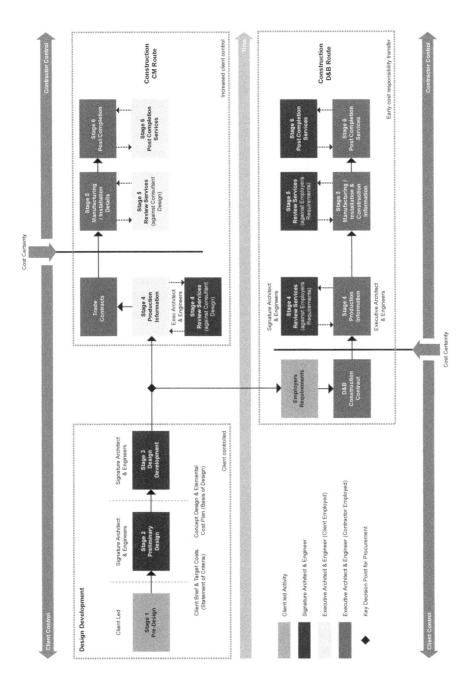

Figure 13.1 Transition points between Client control and Contractor control.

let. This leaves the client with a degree of uncertainty over the cost out-turn, which may be deemed as unacceptable to the client governance. The ability to negotiate and adapt as the individual trade packages are let is likely to enable flexibility and the risk is being taken forward by the client so the main contractor is not required to hedge this risk.

The second option is to transfer early to get an early commitment to an overall cost, but accept that this hands over control to the contractor from the client. If the client at this point has not adequately defined his employer's requirements then he gets what the contractor gives him. The contractor, in taking the earlier risk on cost, may quite reasonably have included a price for the risk associated with this transfer of single-point responsibility. The client cannot expect to stay in control of detailed issues, having already passed responsibility across to the contractor.

Long-term supply chain relationships

Few clients are commissioning projects with sufficient frequency to be able to develop meaningful ongoing commitments with the construction supply chain. Construction involves many trades, from earthworks contractors, foundation contractors, structural works, utilities, mechanical and electrical systems, controls and instrumentation, fit out and finishes, transportation installations and other specialist systems, through to commissioning specialists. All of these also involve links with manufacturers, materials and products suppliers.

These relationships can be taken for granted, but are crucial to the success of projects. Good contractors have built these relationships over many years through repeat business with their supply chain on a sequence of projects. This continuity of relationships enables a well-established mutual understanding. There is a mutual interest in not letting each other down as the relationship is more important than just one project.

This obviously works well for projects of an established nature comprising tried and tested materials, construction scope and methodologies. It needs greater attention on projects with novel features or when the contractual basis is set up in a new way that the supply chain may not recognise.

The pain/gain incentivised contract involving a target price and actual costs was great until we realised that the concept was not being passed down by the principal contractor to many of the subcontractors. They were still being appointed on traditional lump-sum contracts. However, on inspection we realised that many in the supply chain could not identify what we were defining as actual costs. Their accounting systems were not sophisticated enough to record incurred costs in such detail. They just knew the overall price that would cover their work and give them a reasonable operating profit.

Creativity comes from product development

The real progress in the construction industry comes from the development of new materials and new products. This stems from the product manufacturers who are consistently seeking to improve the performance of their products in order to maintain a competitive edge. Interestingly it also stems from new emerging firms who have found a niche in the market for which they have created a unique offering.

Many of these emerging firms fit into the SME category (small to medium-sized enterprises). New firms with advanced technology ideas may be linked, directly or indirectly, with research institutions such as universities. Initial research and development may have come from an academic base. However, the big challenge is to translate a good academic concept through the process of commercialisation into a product with a real market. The successful SMEs are the companies that have found a route to market and are able to progressively expand at a rate that they can cope with. They need enough time to take their product through the inception and validation pre-production phase and to establish the means to deliver it in the required quantities to be viable.

Access to the right markets is therefore very important to product manufacturers and SMEs. A mature market tends to have vertically well-integrated structures, so that the major contractors have close established links through their supply chains to the manufacturers, including SMEs. The major contractors can bring in an SME for the appropriate scope by knowing their particular capabilities and by helping them to plan delivery.

An interesting observation is that the degree of vertical integration has an influence on the ability for the broader construction sector to engage with the export market. In the UK the Government is working hard to increase the country's international exports. This includes encouraging SMEs to engage with the export market. However, a fundamental characteristic of our British contractors makes this laudable ambition difficult when compared with many international competitors. UK consultants have been enormously successful internationally. But the reality is that when they work internationally it is almost always in conjunction with non-UK contractors. It is with the internationally-minded German, French, Spanish and increasingly Chinese, South Korean, Turkish and even Brazilian contractors. By comparison British contractors are notably absent or not big-league players internationally.

The SMEs in the delivery chain need the umbrella of the contractors with whom they have established home-market relationships. So it is the German SMEs that the major international German contractors pull in, and so on for other international contractors. By doing so, they achieve a vertically integrated construction sector, to the benefit of their country in the international export markets. This is all supported by their respective national strategies for export markets. In turn this actually makes the UK home market vulnerable to these international competitors, bringing their SMEs with them into UK markets.

The ability to build (through a well-ordered site establishment)

In much the same way that the best clients can achieve an overall project culture ('the way things are done around here') the best contractors achieve a site environment and a site culture that instils a focus on good order and safety. There is an atmosphere on a good site that things are done in an effective and controlled way, with everyone knowing what the rules are.

Any client seeking to appoint a major contractor for their ability to build would be well advised to ask to visit their current sites. Particularly, if it is possible, to visit the site which the proposed leadership team is transferring from.

The best contractors will be able to explain and demonstrate how the site is organised and the thinking behind this. They will be able to describe compliance with programme for the different construction work-fronts and how they have handled unforeseen complications.

The first impression of a site is how access into it is managed and controlled. On good sites, new starters or visitors will be expected to go through an induction and site safety briefing. These will be appropriate for the stage the project has reached and will include providing an awareness of the nature of the current work activities.

General site husbandry also makes an impression. Well-organised sites have controlled routes for vehicular and pedestrian movement. They do not have rubbish and surplus materials left lying around. They have hazardous or flammable items securely stored when not in use. They have plant and equipment that is in good condition and properly tagged for safety inspection. Site safety should not be about having a filing cabinet full of forms; it should be the proactive monitoring to check that safe conditions and working methods are actually being achieved on site. An important aspect of site safety is the security of the site perimeter to prevent unauthorised access, particularly by children who might think the site is intriguing.

Site welfare facilities for the operatives will be to good standards and give the impression that the well-being of those working on site matters. Decent first-aid facilities will be available. Operatives will be wearing good-quality personal protective equipment (hats, high-viz jackets, boots, gloves, safety goggles) displaying who they work for, and giving the positive impression that they have pride in the firm they work for. All this makes sense as not only does it lead to a safer and better controlled site, but it also improves productivity.

The quality of work also benefits. A site that is a mess is not an environment that is conducive to top-quality workmanship. It is interesting to note that other sectors have also picked up on this. For instance, the best car manufacturers have servicing facilities with immaculate conditions. Their car servicing no longer has the aura of greasy mechanics and pools of oil; the servicing space is light and airy, with clean floors and with the aura of computer technology and modern sophistication. The expectation is one of good quality, reliable work. Many modern factories are similar.

Large construction sites can be complex and daunting, with the act of construction resulting in constant change. A route within the site one week may not be there the following week. One important consideration is the time it takes for construction workers to reach their place of work and finding that they have the right materials and equipment when they get there. Productivity can be seriously affected by the lost time of getting through security, walking back and forth across the site, to and from the site canteen. A well-planned site locates access points and welfare facilities to make sense for the workers and to get improved output from them.

All of this might seem rather obvious, but do not take it for granted. Good site husbandry does not happen by chance. The laws of entropy tend towards clutter and chaos rather than order, so to counter these needs strong effective management. The best sites have construction leadership that 'walks the talk'. It will not be a surprise or a rare event to see the most senior contractor director out on the site, taking a hands-on interest in the site conditions, the general attitude of those working, what is going on and the safety and quality being achieved.

One further important consideration is the key question:

'Who has the authority on this site to stop work?'

There is always pressure to make progress and maintain output, so it is often seen as unpopular to stop activities. The quality of work or the preparation for the next work may not be quite up to standard, but maybe it will do. Or the site safety might be slightly less good than it should be, but possibly it is good enough to get away with. Everyone wants to achieve their planned outputs, so nobody really wants to stop to correct something. Therefore the common view is that it will take some authority to actually stop work. However, on the best sites this is not the case. It is clear where the buck stops on quality and the delegated right to stop activity and get work corrected. In the case of safety, everyone has the right to not work in unsafe conditions and the team foremen will be empowered to not let activity continue until it can be done safely.

Communicating what has been planned

There is not much benefit in having the best planning of what needs to happen unless this is communicated to those needing to do the work. It also needs to be recognised that those on site may not all be highly educated, so this communication has to be pitched in a way that makes sense to them. They do not need to know everything, just what is relevant and important for them to be able to do their bit to the required quality and safely.

The best contractors now have advanced computing systems. These can take electronic three-dimensional files of the design plus site conditions and apply the fourth and fifth dimensions of time and cost, so that the work sequence and material quantities can be defined. On large sites, GPS (global positioning systems) or GIS (geographic information systems) might be used to map elements of the works and even the movement of site equipment to control the accuracy of construction alignment.

However, all this planning capability needs to be translated into instructions to the labour force. An established way to achieve this is to start a shift with a 'tool-box talk'. A good tool-box talk provides a briefing to a small construction team, often by their foreman. It might start with a review of what was achieved the day before and the consequences of this as a starting point for today's activities. It will then describe the planned work and how this will be shared between the construction team so they all have an idea of what is expected of them. If anyone is less experienced there will be a check on how they will be assisted. The materials, plant and equipment needed will be discussed. The general site conditions in the context of the planned work will be discussed, and any associated risks or safety hazards identified. Working in confined spaces or in live operational areas will be given special consideration and need particular methods for compliance. Tool-box talks will routinely include an update or refresh on safety aspects of the work.

The construction workforce and their unions are inclined to see benefit in ensuring that the project site conditions encourage safety and facilitate output for workers being paid on the basis of work achieved. The local population affected by the construction is likely to also welcome communication and subsequent evidence of promise to maintain good site conditions, and control of site access and the impact on local roads and amenities.

14 Communications policy
How to approach communications

Introduction

The client sets the tone for his project in the way he communicates with the other parties involved. Successful projects are delivered by strong teams with team members that are mutually supportive. The approach to communication can often be included at a high level into the Vision Statement.

The organisational culture of the client body and the visibility of this culture up-front is an important factor; the degree to which it is a hierarchical organisation and the role and authority of the client's project leader. If he is not empowered to make all or certain key decisions without referring to others then this needs to be understood and planned for.

Some issues to consider

Is the operational framework for the project to be one in which others will be informed of what they need to know and then be given a clear remit to just get on and do it? Or is the style to be one of a greater communication of context to those involved so they can see and appreciate the bigger picture?

- How much energy and investment is this worth?
- Does the culture of the various organisations involved in the project matter or need consideration? Be wary of bringing incompatible organisations into the project team, or if electing to do so should the different organisations be encouraged to play to their relative strengths to the benefit of the project?
- Is too much communication both time-consuming and confusing, with the danger of multiple parties all believing they have a say?
- Is the intention to generate an atmosphere of openness and trust, or that of a 'need to know' culture?
- How are errors or mistakes to be handled – is the emphasis on searching for the guilty, in order to assign blame and consequential costs, or is it on how best to collectively achieve a workaround and recovery plan?
- How is reward and praise encouraged? Is there a plan to recognise or incentivise innovation that leads to improved performance?

- How is communication with the outside world handled, and to whose benefit? Does the project need an external identity and profile to make it real and to attract good people, or will this just set it up to be shot at? Communicate with the press and media not when you have an agreed intent, but when you have delivered a result.
- All major projects encounter difficulties so plan for this and have a pre-agreed media communication strategy, with staff trained so as not to be exposed or caught off guard.
- The 'needs of many' balanced against 'interests of a few' – many large projects have their opponents so there should be a balanced understanding within the project team to help avoid the project from being derailed.
- Are team members allowed to publicise and gain reputational benefit from their involvement in the project, or even some of the clever developments that they may have contributed to that might have wider interest and application? In which case, how is confidentiality to be managed and is there an understanding of why this is appropriate?
- Internally there need to be agreed levels of communication, appropriate meeting schedules and defined formats for reports, key performance indicator (KPI) results and the distribution of these.

Avoid superficiality and encourage trust

Say what you mean and mean what you say – *Walk the talk*.

Set targets that matter and explain the context for these – then let the team work out how to achieve them so that they have buy-in and ownership of the objective.

Above all, seek to establish an environment of 'no surprises'. It is better to have problems identified early so that there is a chance of collective discussion on how to overcome them. If the atmosphere of the project is one in which the communication of bad news is a search for someone to blame, then the real situation may be suppressed until it is too late to mitigate.

When to communicate externally

It is often tempting to make public announcements when the project is first established and its objectives have been set in outline. There may be political reasons for making such public commitments.

However, this then invites the press and media to seek opportunities to contest the stated ambitions. Their business is about getting multiple stories – first the public announcement, second the opportunity to print challenges, in the expectation of further printing of rebuffs and counter challenges or of other copy provided by interested third parties without direct responsibility. All of this then needs control by the project team in order to try to maintain the desired image of the project.

It may be safer to hold off making announcements until they are unequivocal achievements.

A separate consideration is that local press and media are always looking for stories and on occasions have the challenge of a 'no news' day. Once under construction, large projects of significance in their patch will almost always have something interesting happening that could be newsworthy. The project can therefore be a 'no news' stopgap. Project teams can build a positive relationship of mutual benefit with the local press or media. The project can help the press to fill otherwise 'no news' days by giving them access to interesting activities. The payback for this proactive engagement will be the press being less inclined to portray the project negatively.

15 External interfaces and social responsibilities

How to handle the interfaces with stakeholders

Introduction

The key issue relating to stakeholders is the degree to which the client is in control of the project's destiny or not. On large projects there can be many external interfaces with those who have a stake or interest. In the broader context the project has an interface with society. Clarity is needed on the many stakeholders, their remits and authority to have an influence. Early engagement in the right way with the stakeholders and external interfaces is crucial if the project is to proceed smoothly. These groups are now well aware of their powers and can use these to resist projects if they have not been appropriately consulted at the right stages.

Some issues to consider

There are different categories of stakeholder and the client needs to identify these and then determine how they are to be managed and by which members of the project team.

Some may be sponsors or co-funders, so have a direct influence on the client decisions, and there is a need to have their 'buy in'. The client needs to manage these closely, but may require information presented by the project team in a specific way to assist this process.

Some may have powers of approval, so need to be involved and consulted. It is usually good to take along such stakeholders in the key decision-making that relates to their area of authority so they feel consulted and that their views are being considered. There may be specific members of the project team who have the best expertise and credibility to handle particular authorities, though the client may need to retain responsibility since these team members may not be able to guarantee achieving approvals.

Some may have the right to be consulted, though they may actually not have any powers to directly influence the project other than by lobbying others who do have powers of approval. Consultations need to be handled by those best able to communicate effectively and at the right level, without patronising or taking the audience for granted. And ideally if there are good ideas that are suggested by the consultees, there is the ability to constructively take these on board and report back.

Some may be affected by the project in such a way that the project has a direct influence on their livelihood or standard of living. If alienated, they can become protest groups with a motive to disrupt or obstruct the project, or to spread negative publicity against the project. These need to be handled with clarity and authority by those empowered within the project to fairly and properly manage both expectations and the outcome, within the powers and authority of the project. If the project offers to mitigate some of the potential consequences then these promises need to be honoured and the client needs to accept and support this.

Some may be end-users so have a stake in the quality and effectiveness of the project. It is always disappointing to complete a project and then immediately have the user say that there are aspects of it that do not really work as well as they should and that with some prior consultation a much better solution could have been achieved.

Benefits to society

Leading organisations are increasingly aware of their role and responsibilities to the society in which they operate, and in which they can be a force for good. Indeed for many clients, particularly in the public sector, being a 'force for good' becomes part of the strategic intent. In addition to leaving a construction project as their legacy, they can also maximise the benefit to society of the footprint of their construction operations.

Social responsibility covers

- Health and safety
- Environment
- Quality of life
- Employment opportunities
- Equality and diversity
- Access and inclusion

Health and safety obligations

The safety, health and environmental aspects of construction projects are all governed by primary legislation, case law and tort. Clients must understand they have specific legal responsibilities under the Construction Design Management (CDM) regulations, which make them accountable for the impact the project has on health and safety. Principally, they must appoint a competent CDM coordinator; ensure adequacy of information, management arrangements, time and resource to allow safe delivery. In the UK this is as follows:

- Safety – Health and Safety at Work Act with its duty of care

- Health – Health and Safety at Work Act and extension of occupational health into the construction industry
- Environmental – Control of Pollution Act

It makes sense for clients to develop, articulate and believe in a strong safety culture, instilled into people's behaviours, and then give a clear indication of expectations in line with this culture – and of their willingness to pay to achieve it. They have a genuine level of belief. The converse is also invariably true, resulting in a cost-cutting race to the bottom. But without strong leadership, the good practice can get crowded out by supply-chain competition – by a member of the supply chain being driven by commercial short-sightedness and being prepared to take the morally repugnant approach of cutting corners on what are ultimately moral issues.

It is useful for clients to address design for safety early for it to have the best effect. Clients involve specialists to provide a valuable input into safety of construction methods and long-term operation and maintenance. In planning for safety, good clients are the guiding mind, able to set as the highest priority the safe execution of all aspects of the project. They encourage a behavioural approach while retaining rules to enable policing, and facilitate peer pressure. They may also consider using accident frequency rates as a part of the contractual payment process.

To reduce accidents, occupational health screening can also be encouraged. Clients can support this on their projects by investing in health professionals to improve productivity and commitment, who use screening to give a message to the whole workforce that they are important. Screening and the presence of a health professional on site beneficially raises the issue of health across a project.

The best clients set standards and boundaries of acceptability tighter than those required as minimums by legislation.

> *It is worth remembering that the highest standard you have the right to expect from others is the lowest standard you exhibit yourself. In health and safety this responsibility to lead by example is paramount.*

Environmental obligations

Environmental impact assessments are commonly required as part of the planning legislation for new projects. These require clients to minimise adverse impact on the environment during both construction and operation phases. Legislation obliges clients to comply with the Control of Pollution Act.

Independent validation is provided by the Civil Engineering Environmental Quality Assessment and Award Scheme (CEEQUAL), which can also be used to improve the environmental and sustainability performance of the project.

- *Waste* – Clients should understand the relevant waste legislation and the issues behind it. For example, site waste management plans place a clear responsibility on clients to initiate excellence by eliminating waste in design.

- *Recycling* – Clients usefully encourage the maximum use of recycled materials by avoiding over-prescriptive specification and motivating the supply chain to increase its use of recycling.
- *Environmental surveys* – Most projects require extensive environmental surveys. Good clients investigate these issues early, as some surveys may need to be done years in advance of the project starting on site.

Quality-of-life considerations

Quality-of-life issues embrace both the project team and the local community. Genuine care for the local community benefits the community and the project. It often pays for clients to encourage a user-friendly attitude from the project team towards the community, placing emphasis on environmental measures which affect quality of life, such as noise, dust and traffic. This also engenders good relations with environmental health officers. Clients can use local meetings, press releases, literature mailshots, local radio and so on to generate confidence within the community. This leads to fewer complaints and minimises any negative press. Clients can also increasingly embrace new technology – such as blogs, text messaging and RSS feeds – to communicate across construction project teams and with the local community. Many communities use local networking sites on the internet – engaging with these websites to keep people informed of construction activity is valued, and at minimal cost.

It is worthwhile for clients to also consider quality of life of the project team. For example, a restriction on working hours can result in increased productivity. After six weeks, one project demonstrated that 40 hours a week gave better construction outputs than 60 hours a week.

> *It was common for our industry sector to work with two twelve-hour shifts per day. However, when we realised that the workforce were often spending two hours each side of their shift getting to and from work and then still needing to get themselves provisioned, fed, etc. we were getting to the point where they were in many cases not having much more than four hours sleep per night. Not surprisingly the consequence of this was fatigue. When we switched to three shifts of eight hours they were having a much more balanced life, were no longer so fatigued and we got improved productivity and safety. The improved productivity actually ended up saving us cost, so everyone benefitted.*

Equality and diversity

Many equality and diversity issues are enshrined in legislation, with prohibitions against discrimination. The best clients are passionate about these issues and able to mentor and encourage their supply chain about the benefits of diversity, to set the vision (e.g. local employment or training to reskill local people) and then to engage with the supply chain and encourage them. They recognise that diversity or a lack of bias within a workforce is a source of strength, but based on meritocracy rather than imposing quotas.

Enlightened clients also encourage their supply chain to consider bias awareness, to avoid a systemic lack of diversity in management processes by ensuring there is no inherent bias in selection for training and development, or reward and recognition. If job advertisements are overtly diverse, this approach can easily be maintained through the selection process – which is preferable to applying diversity late in the process.

Local employment, access and inclusion

Major construction projects tend to have an overall positive benefit and uplift to the local communities. The projects tend to facilitate economic advancement and growth, with major new infrastructure being an enabler for other economic activity. All of this tends to mean that there is a multiplier benefit which is several factors more than just those directly employed in the construction. They may also up-skill those associated with the project so they can in turn apply new expertise on projects and opportunities elsewhere. These can be powerful factors in gaining a willingness for new projects from the communities affected. This obviously apples most when the jobs created are local and not just imported for the short duration of the construction phase.

These are good reasons for clients to promote access and inclusion. The best foster positive links with local schools, including site visits, which all generates interest, involvement and ultimately employment in the industry.

16 Operation, maintenance and legacy
The post-completion and legacy objectives

Introduction

The completeness and effectiveness of a project is as much about its performance in use as it is about the delivery stage. The client has a need and a responsibility to consider and define what is expected for the post-construction activities of commissioning, operation and maintenance. Increasingly there is also a need to consider and define approaches to legacy values and sustainability. Issues to consider include:

- The need for the client to determine whether there is continuity between the project creation phase, putting it into use, operating it and eventual decommissioning.
- Specifying for safe operation and maintenance.
- Consideration of the allocation between capital expenditure and operational expenditure in the context of design life and allocation of responsibilities.
- Aspirations for legacy value, design life and sustainability.
- Planning for decommissioning.
- The importance of planning and establishing project databases and information management platforms.

Planning ahead

Construction projects are generally 'enabling', whether through construction of a building that will provide user facilities, or of infrastructure that will help enable society to function. It can be argued that nothing is of greater legacy value to society than investment today in good infrastructure for the benefit of future generations.

Therefore how the project is received will ultimately be judged by the users, or the population at large, and the values they attribute to it. Clearly cost and programme matter, and need to be responsibly managed against targets, but the success of the project will, in the long term, be judged for its quality and performance in use.

Great projects are ones that capture the imagination and bring delight and technical elegance. Ones that are delivered to the lowest common denominator, driven by delivery managers who have little understanding of longer-term value, may deliver the required utility and this may be exactly what the client requires, but are unlikely to be ranked in the category of great.

So there is an early need as a part of the project inception to plan ahead for performance in use and for the client to be clear about the aspirations for legacy value.

In doing so the ultimate client organisation may need to consider the roles and interests of the various intermediaries if they are to be assigned some of the client function and responsibilities:

- Is the client for the project acting as an agency of the ultimate client for the construction stages only?
- Is the client deemed successful if he can achieve pre-lets 'off plan' and therefore has the goal to get the minimum possible standard to the point where he can fulfil his obligations and take his revenue from the project? Does how the project performs in use matter to him?
- Does the project involve one client body to get the project built and another to fulfil user fit-out requirements? If so, are the interfaces for these understood and defined? What might represent completion to one party may not be completion to the other, or to the ultimate users.
- Is it clear who will be responsible for running and maintaining the facilities, and will they be involved in agreeing to the initial specifications and to the commissioning and hand-over acceptance? In this context, it may be worth imposing some standardisation of components to support future spares and maintenance regimes, even if the overall characteristics of parts of the project need to reflect site-specific criteria.
- Does the client understand the sophistication of what he is buying and understand how to operate and maintain it? Is the new project adopting technologies and operating criteria that are a step-change from the experience and expertise of the client's existing operations and maintenance team and is it realistic to retrain them to meet these new expectations?
- Does the project programme require a phased hand-over and, if so, how is the commissioning of sub-sections of the overall systems allowed for? In many large infrastructure projects there is a need to begin to get revenue streams so that these can help offset the overall financing of the total project, but this complicates commissioning.
- If the client is a 'one-off' for the project, will the client organisation continue to exist for the management and operation of the completed project, or is there a need for a planned transfer to another client body?
- Operating and maintenance instructions, together with acceptance records, as-built records and lists of defects, are normally requirements of the construction contracts. Sometimes this is extended to RAMS (reliability, availability, maintainability and safety/serviceability) databases that are required to be compiled by the construction team. However if there is not an

agreed body with a responsibility to receive, take ownership and understand how to access these often complex documents, there can be a disconnect between seller and buyer.

- Will contracts for maintenance be linked to the contracts for the original supply and installation? The best time to competitively buy maintenance agreements is when suppliers/contractors are still competitively bidding for the installation work. To negotiate with them afterwards leaves the client in a much weaker position.

- A further consideration is that a supplier who has already been committed to a maintenance contract (or at least the client holding the option of one) is likely to put more care and attention into getting the quality of the original installation work right. There will be a vested interest in achieving good quality and high performance if this reduces his future maintenance costs.

There is an adage – *'Plan to finish the job the day you start it.'* So, all of the above points are worth early consideration and planning.

To bring these together, the project programme needs to be more than just the construction timetable. The time needed to properly commission and hand over a complex project can be considerable and this needs to be fully integrated within the project programme from the outset if there is to be a proper under-standing of how to finish the project. Equally, the consequences of phased handovers or the need for separate client or concessionaire fit-out activities need to be planned for.

In many modern infrastructure projects it is not just the physical construction but the customer interfaces, public facilities and information systems that need to be in place for the project to be fully finished and operational. These need to be planned as an integral part of the project programme.

In some large infrastructure projects, it may be that a phased handover is planned in order to gain customer/market feedback that will inform the final definition and specification of subsequent phases.

Does environmental sustainability matter?

Whilst the campaigners now try hard to assert that climate change and global warming is established as being attributed to human influence (using IPCC 2013 and other sources), this may not in itself provide a compelling reason for clients to embrace the carbon-reduction sustainability agenda.

For most clients the agendas that are most worth using are those of 'cost and economics' and 'quality of life', and linking these to the influence on ecosystems and socio-systems.

Carbon (CO_2 = carbon dioxide) represents an output from energy consumption, so less carbon means less energy consumption; energy costs money, so less energy costs less money. Therefore the right carbon savings, viewed in the right way in terms of ecosystems, achieve cost savings. The drivers can therefore be economics and cost reduction rather than environmental, in order to get a beneficial outcome. It can be a virtuous circle.

The crucial challenge in this is having the right ecosystem – one that properly includes not only $CapCO_2$ (capital and construction stage of the asset) and $OpCO_2$ (operational stage for the asset) but also $UserCO_2$ (user efficiency of the asset amenity). It is this complete picture that achieves the true benefit in cost savings to, say, 'UK plc' in terms of our balance of trade and jobs, and by doing so we get the best overall reduction in CO_2. So the more we are able to articulate this and illustrate this overall benefit in our schemes, the better. Our major construction projects should seek to be a market leader in this aspect.

Serious thought needs to be given to how best to articulate this. On new transport links such as HS2 (high-speed rail project) it does not work to try to take carbon savings from other existing systems as you end up 'comparing apples with oranges'. Any new system of infrastructure that involves additional capacity can be looked at to assess the carbon footprint of that new capacity using the new infrastructure, and then the carbon footprint of the old infrastructure if that new capacity had to be imposed onto it. So this is not 'robbing Peter to pay Paul' but a better and simpler direct carbon comparison for the additional new capacity. It should not matter if this new capacity is demand forecast. Good, new, modern infrastructure can deliver new capacity with minimal environmental impact using NEWT (not environmentally worse than) principles, so the net impact is not worse than existed already.

On the 'quality of life' side of the debate, it is difficult to create a real correlation between individual projects and global factors such as the level of greenhouse gas concentration in the atmosphere. The factors are all too complex and subject to influences way beyond any project's control. It may be more useful to consider the more immediate human scale in terms of socio-systems, such as air pollution in cities. No-one wants infrastructure to contribute to the smog and pea-soupers of the 1950s. Even China is now on a mission to de-pollute its cities.

So a more tangible mission should be to:

- reduce pollution that can influence the quality of life and health, and
- do so in a way that saves overall cost by requiring less energy overall.

These two objectives are totally mutually inclusive. Clients should be helped to embrace them.

The term 'sustainability' can either be misused or interpreted in various ways to suit individual agendas. 'Mitigation' (of contributing factors) versus 'adaptation' (to impact of change) and the use of concepts such as efficiency, security, resilience and durability can be a better way to approach sustainability.

Impact versus resilience

Considering two sides of the same discussion can be a powerful way to help articulate the question *'What could success look like?'* and to help establish a way to proactively contribute to the future through major infrastructure projects:

- The project having an impact on environment and society (and agreeing compensatory measures)
  - client responsible for impact
- General climate change and increasing volatility having an impact on the project (and agreeing how to allow for this)
  - manifests as a risk to client

Major projects are rightly required to go through elaborate environmental assessments to demonstrate that they do not have a detrimental impact on the environment. In other words, the project has to demonstrate it does not impact negatively on external factors. However, this is only one side of the equation. Increasingly, there is a need to address the equation from the other direction. If a major investment is being made into new infrastructure, should it not be equally relevant and worth investing in the consideration of how might external factors impact on the functionality of the project? It is not smart to invest in infrastructure that is not adequately resilient. Examples to consider would obviously include the likes of extreme climate events, shifts in demographics and capacity through new technologies.

It is therefore worth considering supplementing an Environmental Impact Assessment with a parallel Resilience Assessment. This may be even more powerful and important as a part of the project's business case if there is a societal need and dependence on the completed project remaining functional in extreme events.

Figure 16.1 Mitigation and Resilience as complementary objectives.

Operating requirements definition

An experienced client may know that if specified facilities are constructed and properly commissioned then he will be able to implement and achieve his operating requirements. In such cases the client effectively takes responsibility for the performance in use, so the responsibilities of the project delivery team (designers, constructors) effectively end with project hand-over.

In other cases the client may take a different approach. He may know the minimum operational performance that will satisfy his business case, but be open-minded or not know how to define the facilities to deliver this. In which case the client can set a performance requirement and require the project team to determine the specification and execute the work in order to meet this required performance.

These two approaches are fundamentally different and can influence the approach to the project, its procurement and the means by which the parties involved receive payment.

Traditional construction contracts awarded following a design process based on a predetermined client brief tend to suit the first of the above approaches. Design-build-finance-operate contracts tend to be more aligned with the second approach.

One thing that needs to be considered is – *What you measure is what you get.*

If the project sets fairly general requirements for quality then the client has the opportunity to apply this generally in judging the achieved results. By comparison, if the client has prepared a very detailed specification to outline exactly what is required, then he cannot be critical of the project team if the result does not meet some other not-specified criteria or more general undefined aspiration or expectation.

The above points can be applied to the whole project or to specific component parts of it.

In-built obsolescence

The business case for the project will establish the timeframe for the investment and return on this.

The level of specification and the quality of the project can then be matched to suit this in order to optimise return on investment and to achieve the right balance between CapEx (capital expenditure) and OpEx (operational expenditure). If the plan is for a clear handover from the construction team to an operating team then the interface between CapEx and OpEx needs to be unambiguously defined. This is less critical when the same organisation is responsible for both.

Different components of the project will have different life expectancies so care is needed when talking about 'design life'. A project may have a 120-year design life to match with the overall business case. However it is clearly impossible and unnecessary for every single component to be designed and specified to last for this 120 years. So consideration needs to be given to defining what is meant by project design life, the influence of planned maintenance on functionality in use and the need for some redundancy or over-capacity to enable outages.

A further factor may be the cost and effort involved in allowing for future flexibility or adaptability to be able to modify the utilisation, and thus avoid obsolescence. In general, design flexibility can be expensive as it is seldom possible to be able fully to anticipate future requirements and to allow for these up front. Often the ability to be able to adapt is more useful and less expensive in terms of up-front investment.

Decommissioning

One of the great challenges in a sustainable environment is that projects should have no long-term detrimental impact. Picking up on the rightful ambition that

We should not leave the planet in a worse condition than we inherited it in.

Clients clearly have a prime responsibility for this. Their business case and client brief for projects should therefore include not only the cost of building and operating the project but also some consideration of safe and environmentally responsible decommissioning. Materials should be specified and designed to satisfy these whole-project objectives.

Experience shows that most materials can be recycled or reprocessed once separated out. Some can be crushed or reused without major treatment and are reasonably inert. Others need specialist reprocessing that can only be achieved once transported to suitable facilities. Therefore, increasingly, it is the ability to be able to reasonably effectively sort and separate materials that most influences the decommissioning. Design should take benefit of recycled materials from previous decommissioned projects and in turn facilitate future recycling.

Legacy and sustainability

Increasingly it is important to deliver projects that have clearly defined legacy values and which have strong sustainability credentials. But whilst there is much talk of sustainability, the degree to which this is hype or whether it can really be measured to know what has actually been achieved is often less easy to gauge.

The sustainability agenda offers an opportunity for the client to demonstrate leadership. For example, it is often more expensive to use sustainably sourced materials, so if the supply chain's driver is commercial, it is likely that sustainability and best value will be lost in a race to the lowest price. But by considering whole-life costing, and looking at sustainability in both construction and operation, choices in design and construction can improve the long-term sustainability of the project.

Informed clients consider whole-life costing to improve the long-term sustainability of a project. They consider sustainable options during the design as innovations, and new technologies are best considered by the integrated team at an early stage when defining specifications.

Clients are increasingly asking to be able to benchmark what is being aimed at and to have some way of measuring the outcomes and to be able to position their projects relative to other projects. If a client has invested significantly in aiming to achieve particularly high sustainability standards then they like to be able to define and describe this. There are nationally and internationally recognised standards for registering and gaining certification for aspects of sustainability performance (e.g. CEEQUAL, LEED, BREEAM). If the client wants his project to seek such certification then this decision needs to be taken early and the standard sought should be established at the outset, together with the level of attainment (e.g. silver, gold, platinum) within this standard.

Like quality or environmental standards, there is often a need for a degree of independent audit and monitoring associated with sustainability certification. It may, therefore, be appropriate for the client to appoint specialists at an early stage to advise on this and to set up the sustainability objectives and monitoring in order to gain certification.

Travel is a key issue for project sustainability. Video conferencing has an invaluable place, but good clients recognise that relationships need to be started and nurtured by face-to-face meetings, with video conferencing being used once the relationship has developed.

It has been said before, but undoubtedly

> 'One of the greatest legacies one generation can leave to their successors is the excellent infrastructure of well-built resilient construction projects that are positive enablers for our complex socio- and economic- and environmental-systems.'

Good projects make a commitment to contribute to this and to capture the associated thinking.

Project databases and information management

Projects often have many parties involved, either as contributors to the design and construction, or as operators, users and maintainers. The way in which data and information is assembled and used is therefore an important aspect of controlling the project out-turn and whether it is successful.

Good data management can provide consistency, help the visibility and control of interfaces, enable a detailed audit trail of the decision-making and accountability processes and set up a basis for as-built records and maintenance schedules. It can also provide ongoing feedback of performance in use. Design can be assembled as a 3D model, 4D when linked into construction planning and linked into GIS (geographic information systems) or BIM (building information modelling) databases linked to records of approvals and performance. In turn this database will assist safe operation and facilitate decommissioning.

It makes enormous sense for clients to ensure full records are kept of the project construction, and that non-conformances are methodically closed-out. Records should include legacy photos, and any automated reporting systems or GIS/BIM. Long-term use in service, modification and decommissioning are also important; energy costs and recycling have been made more prominent by increases in the price of fossil fuels. This emphasis has initiated a focus within construction on the availability of information on sustainability, through reducing energy costs and increased recycling, which has now become entrenched in the thinking of many clients.

Establishing the project database is therefore an activity that should be planned early as part of the project inception. This needs a plan in its own right to determine the degree to which a single integrated project data platform is to be used and the levels or hierarchy for project information such as programmes, cost plans and risk registers. A further factor is to set up security processes so that certain data can readily be widely accessed, while another has protected limited access for commercial confidentiality. Evolving design and survey data may also need to have access linked to status so that interfaces can be properly managed and controlled to prevent the adoption by others of information that has not yet been properly approved as the agreed design basis. Modern contracts such as the NEC now also have aligned information-management systems to help achieve tracking of instructions and acceptance.

Clients may have established in-house management procedures for data and information, in which case they need to specify how other contributors to their project are expected to interface with these. Alternatively, the client may need to engage specialists or assign to a member of the project team responsibility for defining, assembling and managing databases.

17 Conclusions
Author's wrap-up summary

Some observations relating to style

Through my involvement in many different projects, it has been interesting to observe and reflect on some project relationships, particularly if there are different cultural origins involved.

There was one project on which the client almost always said "Yes" to ideas I presented. However, I began to notice that despite having said yes not all that much was actually progressing. Certainly not at the pace that might have been expected. The client was from the Far East. Mentioning to him that he had said yes to something that was not happening, he looked slightly surprised. Thinking, he responded that by saying yes he was acknowledging he had heard and understood me, but not that he was instructing it to go ahead. Afterwards we had a much clearer understanding and better communication.

On another project in Seville, Spain, we had flown down for an important meeting. We spent the morning presenting all of our work and proposals. But then the senior client said he wanted to invite us all to go for a good lunch. I was busy trying to say that we needed to reach some decisions as the client continued to head us out to lunch. After a long, very pleasant lunch with me still wondering what we had achieved, he got up came around the table, shook my hand and announced that he would like to say he agreed to everything we had proposed. It was more like a personal gift, but it turned out to also be a very clear project commitment.

The contrast in style could not have been more marked than on a major project in Germany. At this we could not get up and leave project meetings until the client's project manager had written out the minutes and we had all been through every word to agree both to them and to the assignment and acceptance of responsibilities for all of the actions. The thought of a good lunch before this would not have been contemplated.

In some cultures only the most senior representative will speak and voice opinions to the meeting, with the more junior respectfully demurring. In others everyone with the view they have something relevant to contribute feels free to do so. Meetings with a mix of background cultures can need careful handling to achieve proper understanding. The norm in some parts of the world is to have open bartering and this is not seen as unprofessional. For some cultures respect is shown by making the effort just to go and have a coffee to exchange pleasantries with the senior client.

On some projects, meetings are unambiguously structured for one side to make their case to the other side. Whereas others are structured so representatives (client and project team) at one level work as a collective to report up to a joint group of more senior representatives. The dynamic of these two styles of meeting is very different.

It always helps to have an awareness of the influences at play.

Final thoughts

I hope you have found this book a useful read, or at least have found sections of it relevant to your areas of interest.

It is a compendium of thoughts and observations across the full range of project activities. It is therefore not an attempt to set out a single magic formula or approach that can be always the right answer. The chapter headings are all topics that need in some way to be considered and addressed. In doing so there needs to be a guiding mind, and the client should be central to this. After all it is the client's project. Everyone else is there to help achieve the client's ambitions.

I have written this as a design-based professional engineer – as an experienced practitioner, not as an academic, researcher, bureaucrat or manager. What drives me is the excitement of being involved in directly contributing and seeing projects conceived, designed and constructed well.

Good projects are where those involved all have specific skills to be able to make a positive contribution and provide an important input. Then the whole can be greater than the sum of its parts. On good projects there is not a need to fill them with individuals whose only real role is to manage others – particularly if these managers do not really understand what they are managing, the consequences of the interfaces they create and the need for associated processes to imply control. It is also nice to avoid engaging teams simply to unnecessarily man-mark others. This dilutes responsibility, adds cost and time.

Win–win is the ambition. It is great to empower by setting out the aims and the art of the possible. Then let those involved use creativity and commitment to rise to the challenge.

Some obvious conclusions would include:

- Know what you want to achieve before you ask others to do it
- Planning up front is more sensible than planning on the run
- Once in the implementation phase, generally time costs money, so one of the best ways to control the out-turn cost is to keep a control on programme
- Scope creep and new ideas at the wrong time are one of the easiest ways to lose control

- Spend time and effort with those you do want to work with, rather than putting too much effort into eliminating those you do not
- Most people want to be associated with success and are willing to put in considerable personal effort if they believe it can be achieved
- No surprises
- Treat others as you would like others to treat you

Index